T0136080

About Island Press

Since 1984, the nonprofit Island Press has been stimulating, shaping, and communicating the ideas that are essential for solving environmental problems worldwide. With more than 800 titles in print and some 40 new releases each year, we are the nation's leading publisher on environmental issues. We identify innovative thinkers and emerging trends in the environmental field. We work with world-renowned experts and authors to develop cross-disciplinary solutions to environmental challenges.

Island Press designs and implements coordinated book publication campaigns in order to communicate our critical messages in print, in person, and online using the latest technologies, programs, and the media. Our goal: to reach targeted audiences—scientists, policymakers, environmental advocates, the media, and concerned citizens—who can and will take action to protect the plants and animals that enrich our world, the ecosystems we need to survive, the water we drink, and the air we breathe.

Island Press gratefully acknowledges the support of its work by the Agua Fund, Inc., The Margaret A. Cargill Foundation, Betsy and Jesse Fink Foundation, The William and Flora Hewlett Foundation, The Kresge Foundation, The Forrest and Frances Lattner Foundation, The Andrew W. Mellon Foundation, The Curtis and Edith Munson Foundation, The Overbrook Foundation, The David and Lucile Packard Foundation, The Summit Foundation, Trust for Architectural Easements, The Winslow Foundation, and other generous donors.

The opinions expressed in this book are those of the author(s) and do not necessarily reflect the views of our donors.

SOCIETY FOR ECOLOGICAL RESTORATION

The Science and Practice of Ecological Restoration
Editorial Board
James Aronson, EDITOR
Karen D. Holl, ASSOCIATE EDITOR
Donald A. Falk, Richard J. Hobbs, Margaret A. Palmer

A complete list of titles in this series can be found in the back of this book.

The Society for Ecological Restoration (SER) is an international nonprofit organization whose mission is to promote ecological restoration as a means to sustaining the diversity of life on Earth and reestablishing an ecologically healthy relationship between nature and culture. Since its incorporation in 1988, SER has been promoting the science and practice of ecological restoration around the world through its publications, conferences, and chapters.

SER is a rapidly growing community of restoration ecologists and ecological restoration practitioners dedicated to developing science-based restoration practices around the globe. With members in more than fifty countries and all fifty US states, SER is the world's leading restoration organization. For more information or to become a member, e-mail us at info@ser.org, or visit our website at www.ser.org.

INTELLIGENT TINKERING

Intelligent Tinkering

Bridging the Gap between
Science and Practice

Robert J. Cabin

ISLANDPRESS
Washington | Covelo | London

Typesetting and text design by Karen Wenk
Printed using Electra

Library of Congress Cataloging-in-Publication Data

Cabin, Robert J.
  Intelligent tinkering : bridging the gap between science and practice / Robert J. Cabin.
    p. cm.
  Includes bibliographical references and index.
  ISBN-13: 978-1-59726-963-6 (cloth : alk. paper)
  ISBN-10: 1-59726-963-8 (cloth : alk. paper)
  ISBN-13: 978-1-59726-964-3 (pbk. : alk. paper)
  ISBN-10: 1-59726-964-6 (pbk. : alk. paper) 1. Restoration ecology. 2. Experiential research.
I. Title.
  QH541.15.R45C33 2011
  333.73'153—dc22

                                                                                          2011005767

Printed on recycled, acid-free paper

Manufactured in the United States of America
10 9 8 7 6 5 4 3 2 1

Keywords: Island Press, science, restoration practice, Aldo Leopold, restoration ecology, ecological restoration, science-practice gap, conservation biology, Hawaiian ecology, tropical dry forests, practical relevance of science, restoration practitioners, trial-and-error experimentation, restoration philosophy, limits of reductionism, alternative research paradigms.

To Anne, Ellen, and Paul.
Of all the gifts I've been fortunate enough to receive,
your love and laughter have been the best by far.

CONTENTS

ACKNOWLEDGMENTS

I strived to research and write this book as objectively and accurately as possible. However, given the tightly knit, politically charged, and often volatile nature of Hawai'i's environmentally related academic, conservation, practitioner, regulatory, and scientific research communities, in a few instances I deliberately altered or obfuscated some of the factual details surrounding particular people and projects in order to preserve anonymity. I also altered the chronology of a few minor events for the sake of narrative clarity and brevity.

I transcribed all of the substantive quotations in this book from my taped oral interviews; because I promised my subjects anonymity in exchange for candor, I did not name these quoted people except (with their permission) when it was impractical not to identify them (e.g., for the major players in the "intelligent tinkering" case studies in chapter 9). Since I obviously did not record my less formal conversations with friends and colleagues, these quotations necessarily represent my best approximation of what was said.

During my years as a resident of Hawai'i, I had to endure a few too many jet-lagged visitors who expected me (and everyone else who lived and worked on the islands) to drop everything in order to answer their often naïve questions and meet their various and considerable needs before they flew away. Thus, I am especially grateful to all those who patiently answered *my* naïve questions and graciously accommodated *my* needs while I was working on this book; your *aloha* and *mālama* have and continue to mean more to me than you will ever know. These people include Adam Asquith, Donna Ball, Randy Bartlett, Tom Bell, Dave Bender, Wayne Borth, Marie Bruegmann, David Burney, Katie Cassel, Mick Castillo,

Chuck Chimera, Melissa Chimera, Colleen Cole, Susan Cordell, Ellen Coulombe, Julie Denslow, Saara DeWalt, Leilani Durand, Mary Evanson, Kerri Fay, Tim Flynn, Charlotte Forbes, Betsy Gagne, Bill Garnett, Christian Giardina, Jim Glynn, Don Goo, Kawika Goodale, Lisa Hadway, Richard Hanna, Audrey Haraguchi, Eileen Harrington, Roger Harris, Pat Hart, Stephen Hight, Baron Horiuchi, Flint Hughes, Jim Jacobi, Jack Jeffrey, Tracy Johnson, Jordan Jokiel, Boone Kauffman, Kapua Kawelo, Creighton Litton, Rhonda Loh, Lloyd Loope, David Lorence, Art Medeiros, Tami Melton, Theresa Menard, Trae Menard, Nancy Merrill, Alex Michailidis, Matthew Notch, Becky Ostertag, Steve Perlman, Lyman Perry, Karen Poiana, Linda Pratt, Peter Raven, Don Reeser, Joby Rohrer, Darren Sandquist, Paul Scowcroft, Phyllis Somers, Hannah Springer, Bryon Stevens, Bill Stormont, Jarrod Thaxton, Mike Tomich, Pat Tummons, Tim Tunison, Alan Urakami, Erica von Allmen, Warren Wagner, Rick Warshauer, Dick Wass, Steve Weller, Chipper Wichman, Haleakahauoli Wichman, and Ken Wood. To the many other knowledgeable, talented, and inspiring people in Hawai'i whom I was unable to catch or never fortunate enough to meet or correspond with, *mahalo* just the same.

I am grateful to the people and institutions that provided much-needed support and guidance during the earliest stages of this project. The 2004 Bread Loaf Writers' Conference, especially my participation in William Kittredge's nonfiction workshop (superbly assisted by Kristin Henderson, Sebastian Matthews, and my fellow students), was especially influential and helpful. The people associated with the Institute for Ethics in Public Life at Plattsburgh State University of New York provided a stimulating and productive environment in which to write and think throughout my time there as a Fellow. As the book developed, Heather Fitzgerald, Ibit Getchell, Eric Pallant, Cherry Racusin, Tom Vandewater, Paddy Woodworth, and students in several classes at Brevard College read one or more chapters and offered much-appreciated encouragement and constructive criticism.

Barbara Dean, my editor at Island Press, provided superb and extensive guidance throughout the latter stages of this project that dramatically improved the book's overall content, quality, and organization; my only regret is that I wish I had been lucky enough to work with her sooner (so does she!). I am grateful to Pat Harris for excellent copy editing and to Erin Johnson for additional editorial assistance at Island Press. I also thank Jack Jeffrey for his beautiful photograph that graces the cover of this book, and Chris Robinson for producing the maps of the Hawaiian Islands.

I am deeply indebted to Andre Clewell and Randall Mitchell for generously reading the entire draft of an earlier version of this manuscript and providing substantial and invaluable expert criticism and support. James Aronson offered insightful and constructive comments on a later draft of the book as a whole and the last chapter in particular.

Finally, for their endless patience and unflagging support, my family deserves the greatest thanks of all, so Anne, Ellen, Paul, Rhea, and Seymour: *Mahalo nui loa!*

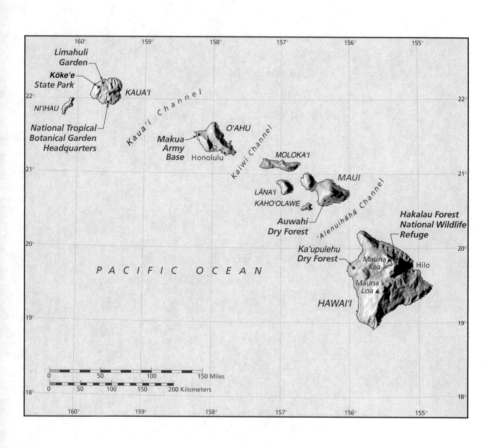

Limahuli
Garden

Kōke'e
State Park

KAUA'I

NI'IHAU

National Tropical
Botanical Garden
Headquarters

Kaua'i Channel

O'AHU

Makua
Army
Base

Honolulu

Kaiwi Channel

MOLOKA'I

MAUI

LĀNA'I

KAHO'OLAWE

Auwahi
Dry Forest

'Alenuihāhā Channel

Hakalau Forest
National Wildlife
Refuge

Ka'upulehu
Dry Forest

Mauna
Kea

Hilo

P A C I F I C   O C E A N

Mauna
Loa

HAWAI'I

0   50   100   150 Miles

0   50   100   150   200 Kilometers

Introduction:
The Science of Restoration Ecology and the Practice of Ecological Restoration

I could see the charred remains of the ghost forest from the highway. One mile below me, the dead trees rose from the lava like giant skeletons. There were many reasons not to walk down there: the steep slope, the intense heat, the dark and foreboding lava, the dense swath of neck-high African fountain grass I would have to fight my way through to reach the 200-year-old lava flow that ran down to the ruined trees. More than all of this, I didn't want to go because I'd been in Hawai'i long enough to visualize the ecological devastation I would see when I got there. But something I could no longer ignore compelled me to go.

I swung my legs over the guardrail, stepped off the highway, and plunged into a sea of dead grass. A prolonged drought on this side of the island had reduced tens of thousands of acres of formerly lush fountain grass (*Pennisetum setaceum*) to a brown wasteland. Head down, I trudged toward the lava as if walking against a strong, waist-deep current. Inside the tunnel of grass, the air felt heavy and smelled like rotten hay. The brittle stems scratched at my bare arms and legs; after ten minutes I felt the familiar sting of sweat trickling into my blood.

When I reached the flow, I could feel the heat radiating from the black rock through the thin soles of my shoes and into my blistered feet. I paused to brush the fountain grass debris off my face, gulp down some water, and look around. My eyes followed the meandering route of the lava flow past the burned forest and all the way down to the sparkling ocean some six miles and 2,000 feet below me. Across the channel, seventy miles northwest from where I stood on the island of Hawai'i (the "Big Island"), East Maui's 10,000-foot Haleakalā Volcano rose majestically out of the sea, and

1

I could just make out the faint outlines of the islands of Kahoʻolawe, Lānaʻi, and Molokaʻi floating on the horizon west of Maui.

I shouldered my pack and set off across the lava for the forest. There are two main kinds of lava in Hawaiʻi: when relatively fluid magma cools, it forms smooth, solid, ropy *pāhoehoe*, while relatively viscous magma forms rough, rubbly, clinker-type *ʻaʻā*. Even though this was a somewhat treacherous *ʻaʻā* flow (falling on this type of lava often results in nasty cuts and gashes), the walking here was much easier and faster than within the fountain grass. When I first began working here as a restoration ecologist for the National Tropical Botanical Garden in 1996, I wore expensive, sturdy hiking boots, but after the *ʻaʻā* destroyed my second pair I gave up on the concept of ankle support and switched to cheap, low-cut sneakers. Eventually I acquired my "*ʻaʻā* legs" and rarely fell except when I let my eyes and mind wander too far from my feet. Fifteen minutes into this hike, when I tripped over a loose piece of lava and nearly stumbled into a jagged ravine, I realized with a jolt that I had been looking at the coast and daydreaming about the ocean. It had been far too long since I'd swum and surfed and snorkeled in those waters.

The lowland, dry, leeward sides of all the main Hawaiian islands were once covered by magnificent forests teeming with strange and beautiful species found nowhere else on Earth. Tens of thousands of brightly colored, fungi-eating snails slithered through the trees and inched their way through the dark underlying leaf litter. Vast flocks of giant flightless geese squawked across the forest understories; dozens of species of finchlike honeycreepers sipped nectar, gobbled insects, and sought shelter from the heat and hungry eagles, hawks, and owls.

Paradoxically, the diversity of Hawaiʻi's primeval dry forests was probably created and maintained by rivers of red-hot molten lava that destroyed everything in their path as they wound their way down the slopes of the volcanoes and into the sea. Before alien species such as fountain grass reached these islands, the native plant communities apparently did not produce enough understory biomass to carry fires much beyond the lava rivers, so the forests on either side of the flows remained more or less intact. Thus, as each wave of new lava cooled and weathered, it was slowly colonized by the species in the adjacent forests. The result of thousands of years of this dynamic cycle was a mosaic of different-aged forests, with different species assemblages growing sometimes literally side by side.

The Hawaiians loved these forests and often chose to live in or near them. Because of the hot, dry climate, many of the trees grew extremely slowly and produced some of the world's hardest woods, which the Hawaiians fashioned into buildings, tools, weapons, and musical instruments. They also made exquisite multicolored capes and helmets containing hundreds of thousands of bird feathers and strung elaborate leis using vines and sweet-smelling flowers.

The first time I walked through a patch of native dry forest containing a grove of *alahe'e* trees in full bloom (*Psydrax odorata*, a member of the Coffee family), I told my native Hawaiian colleague that the light fragrance of these small white flowers seemed to creep mysteriously in and out of my nostrils. He smiled and explained that the Hawaiian word *alahe'e* literally means "to move through the forest like an octopus."

Today we can only imagine what these complex ecosystems looked like and guess at how they worked. Tragically, more than 90 percent of Hawai'i's original dry forests have been destroyed, and many of their most ecologically important species are actually or functionally extinct. For example, most of the native birds and insects that once performed such critical services as flower pollination and seed scarification and dispersal are now gone. Many of the once dominant and culturally important canopy trees are also extinct or exist in only a few small populations of scattered and senescent individuals.

The demise of Hawai'i's dry forests began soon after the Polynesian discovery of these islands around AD 400. Like indigenous people throughout the tropics, these early Hawaiians cleared and burned the dry coastal forests and converted them into cultivated grasslands, agricultural plantations, and thickly settled villages. In 1778, Captain James Cook became the first white man to reach Hawai'i when he accidentally discovered the archipelago while searching for a northwest passage between England and the Orient. Cook's arrival set in motion a chain of events that dramatically accelerated the scope and intensity of habitat destruction and species extinctions throughout the Hawaiian Islands. While the Polynesians had deliberately brought many new species to Hawai'i in their double-hulled sailing canoes (and some stowaways, such as the Polynesian rat, geckos, skinks, and various weeds), their impact was trivial compared with that of the ecological bombs dropped by the Europeans. Thinking the islands deprived of some of God's most useful and important species, Cook and his successors, with the best of intentions, set free cows, sheep, deer, goats, horses, and pigs. Over time, foreigners from around the world unleashed a veritable Pandora's box of ecological wrecking machines, including two more rat species, mongooses (in an infamously ill-advised attempt to control the rats), mosquitoes, ants, and a diverse collection of noxious weeds such as fountain grass.

During relatively rainy periods, when fountain grass greens up and is in full bloom, large sections of the leeward side of the Big Island can look like a lush midwestern prairie. But inevitably the merciless Kona sunshine returns, and the rains disappear for months on end. All that fountain grass dries up and changes from bright green to sickly brown, and the whole landscape looks as if it had been sprayed with Agent Orange. Then all it takes for the whole region to burst into flame like a barn full of dry hay is for somebody to park a hot car on a clump of fountain grass or throw a cigarette out the window.

In contrast to most Hawaiian species, fountain grass originated in an ecosystem (North African savannas) that regularly burned, and consequently it has had thousands of years to evolve mechanisms to cope with and even exploit large-scale fires. I have watched fountain grass rise up from its ashes like a green phoenix after seemingly devastating wildfires: vigorous new shoots quickly appear within the old, burned clumps; seeds germinate en masse; and the emerging seedlings rapidly establish themselves in the favorable postfire environment of increased light and nutrients and decreased plant competition.

The net result of these fires is more fountain grass and less native dry forest. More grass means that during ensuing droughts there will be even greater fuel loads, which in turn will lead to more frequent and widespread fires. This cycle of alien grass, fire, more alien grass, more fire has proven to be the nail in the coffin for dry forests on the Big Island and throughout the tropics as a whole. The reason we don't hear about campaigns to save tropical *dry* forests is that there are now virtually no such forests left to save. If we want at least some semblances of this ecosystem to exist in the future, we'll have to deliberately and painstakingly design, plant, grow, and care for them ourselves.

As I approached the dead trees, I was hot and felt frustrated because I had never seen this forest before it burned. Yet, in a bittersweet way, I was also glad I had not, because even with no personal connection to this place, I found the sight of those scorched trees almost unbearably depressing. This had apparently been one of the best native dry forest remnants left in the entire state, but we would never know which species had lived here or even what the canopy tree, shrub, and understory layers had looked like. We would never be able to collect seeds or cuttings from the gnarled old trees, which had thrived here against all odds for hundreds of years but now were on the very edge of extinction. One more irreplaceable piece of the mysterious Hawaiian dry forest ecosystem puzzle was gone forever, leaving behind only some tantalizing clues in the fading memories of the few remaining people who had seen these trees alive.

It was a miracle that this forest had survived to the last decade of the twentieth century. Its continued existence was probably due to its location within a large *kīpuka*—an island of vegetation surrounded by a sea of barren lava. The wide sheets of *'a'ā* that encapsulated it must have served as both a natural firebreak and a physical barrier to the herds of goats and cattle that roam these lands looking for something to eat within the endless fields of unpalatable fountain grass. Nobody knows for sure how fire finally managed to penetrate this *kīpuka*. Perhaps fountain grass's steady colonization of its surrounding lava shield provided enough fuel for the fire to hopscotch its way in. Perhaps the wind simply blew a clump of burning grass into its interior. Or maybe, as some say, the fire was deliberately set by a disgruntled rancher or bored teenagers.

By the time I finally reached the dead trees, I had seen more than enough to satisfy my curiosity and my conscience. There were no new leaves or shoots on the trees, no regenerating native shrubs or vines, no seedlings or seedpods on the ground. Up close, the blackened trunks looked more like tombstones than ghosts. I could tell I was looking at the corpses of several different kinds of tree, but I could not determine with any confidence which species they were. Although such hard, dense wood takes forever to rot in this parched environment, I knew it would not be long before the last tree toppled over and disappeared in the underlying thicket of rank fountain grass.

I wiped the sweat out of my eyes and looked toward the Kohala Mountains, twenty-five miles to the northeast, but all I saw was mile after mile of fountain grass interspersed with more barren, black, bleak lava flows. The view to the southwest was only marginally less discouraging: while there were still a few scattered bands of native trees poking up here and there, I saw new roads going in and new construction projects going up virtually everywhere. The Big Island's famous Kona coastline to the west was a mixture of raw lava, groves of thorny alien *kiawe* trees (*Prosopis pallida*, or mesquite), and the kind of high-end resorts that rent private pieces of well-stocked paradise for many thousands of dollars a night. Only a few miles away from the *kīpuka*, I spotted the lush greens and glittering, volcano-motif copper clubhouse of Charles Schwab's new $50 million private golf course: apparently he had not found any of Kona's fifteen existing golf courses quite up to par.

I turned away from the sea and the opulence and looked back upslope at the tiny parcels of native trees lining the highway. The North Kona

Dryland Forest Working Group had collectively spent thousands of hours to preserve and restore those forest remnants. We had erected and maintained fences, established perimeter firebreaks, killed and cleared fountain grass and other weeds, poisoned rodents, collected seeds, and propagated and transplanted thousands of native trees, shrubs, and vines. Local groups ranging from elementary school kids to native Hawaiian teenagers to real estate agents had repeatedly donated their time and labor to help with these efforts. Hundreds of people within and beyond the Hawaiian Islands had come to see and study this ecosystem. My own scientific research program had progressed from documenting the demise of these forests to experimenting with promising techniques for restoring them at ever larger spatial scales.

Looking at the fruits of our work from this distance, I felt a wave of optimism sweep over me, and for the first time I truly believed that even this saddest of all the sad Hawaiian ecosystems could be saved. I turned around again and looked at the ruined trees. "We can grow another forest here," I muttered. "We know what to do and how to do it."

As the eminent ecologist, conservationist, and pioneering wilderness advocate Aldo Leopold once observed, those who care about the natural world and are aware of what we have done and are doing to it often live "alone in a world of wounds." Environmentalists are almost always forced to play defense: fighting to maintain and enforce hard-won yet meager environmental regulations, scrambling to halt the construction of the next shopping mall, lobbying to preserve the integrity of our last few crumbs of relatively wild and untrammeled places. Thus, one of the most powerful aspects of ecological restoration is that it offers a rare opportunity to go on the offensive; those who do it usually get to, at least occasionally, enjoy the sweet satisfaction of seeing degraded ecosystems and communities and species reverse course and get better.

On one level, ecological restoration involves a seemingly simple two-step process: (1) identify and remove or mitigate the factors that have created the degradation and (2) recreate the biotic and abiotic conditions that existed in the area before it was degraded. Although in practice this procedure is usually anything but simple, in many cases it is surprisingly effective. And compared with, say, converting a barren lava field in the middle of the ocean into a world-class golf course, ecological restoration can be surprisingly affordable.

On another level, however, doing ecological restoration is much like raising children: just about everyone involved has a strong opinion about

how it should be done, but no one has the wisdom to know or the authority to decree exactly how to do it. (In the same way that many people are
the "perfect parents" until they have children, many are the "perfect
restorationists" until they meet their first real-world project.) Designing
and implementing a restoration program can be an intensely political process that involves diverse and disparate individuals and interest groups.
Myriad technical and scientific questions must be addressed: Should the
dominant alien species be poisoned or manually removed? Should the last
few wild individuals be captured and captively bred or left alone and their
habitat improved? In addition, restoration ecologists frequently grapple
with equally if not more important and difficult philosophical, social, and
economic questions. For example, whose set of values should guide a project, that of expert scientists, the local community, or the region's indigenous people? Who will pay, and who will benefit? Who will give the orders, and who will follow them? How will we know whether or not we have
succeeded, and who gets to be the judge?

The paradoxes of Hawai'i provide a fascinating microcosm in which to
examine the theory and practice of ecological restoration. On one hand,
Hawai'i has been an unmitigated ecological disaster. Despite the fact that
the Hawaiian Islands represent a mere 0.2 percent of the land area of the
United States, three-quarters of all the bird and plant extinctions in America have occurred within this archipelago, and all four Hawaiian counties
now rank in the country's top five counties for federally listed endangered
plant and animal species. Hawai'i also has the worst alien species problems
in the United States, if not the entire world: one can spend days traveling
across the state admiring the islands' "thousand hues of green" and literally never see a single native species.

On the other hand, Hawai'i still has about 12,000 extant species that
exist nowhere else in the world, and more new or "extinct" species are discovered every year. While some hard-boiled conservation biologists believe the United States should practice ecological triage and give up on
these islands, others argue we should put our money where our endangered species are (one-third of all of America's presently threatened and
endangered birds and plants now reside in this single state) and make preserving and restoring Hawai'i's native biodiversity one of our highest conservation priorities.

Hawai'i's biogeographic and socioeconomic paradoxes simultaneously
make these islands breathtakingly unique and intimately connected to the
rest of the world. The Hawaiian archipelago was one of the last places to
be discovered by humans because these are the most isolated islands on

the planet. Yet largely because of this extreme isolation, a few of the non-human immigrants that managed to get there and establish viable populations ultimately evolved into some of the world's most fascinating and unusual species. For example, more than 90 percent of the native flowering plants and 80 percent of the native birds are endemic to the islands. Although Hawai'i's eight major islands lie between latitudes N 18°54′ and N 22°12′ (a geography roughly equivalent to that of Cuba), all of Earth's climates and most of its ecosystems are represented there. And because in modern times people from all over the globe have flocked there, Hawai'i is one of the most culturally and racially diverse places in the world. But while its unsurpassed beauty, delightful climate, and political stability continue to attract the ultrarich and famous, much of Hawai'i remains poor, insular, and reminiscent of a developing country.

For better and worse, Hawai'i became America's fiftieth state in 1959, making it the planet's only tropical region regulated by the full arsenal of US environmental rules and regulations. However, officials attempting to implement and enforce these laws in Hawai'i often face the same kinds of challenges they would encounter in developing tropical countries: the ranks of relatively rich, well-educated, white *haole* (foreign) environmentalists who come to save Hawai'i have not exactly been welcomed at the airports by throngs of lei-bearing locals. Yet if we Americans fail to preserve and restore our only tropical ecosystems, can we continue to lecture the Brazils and Borneos of the world about the importance of saving theirs?

The science of restoration ecology has been called the "acid test" of academic ecology: if we really understand how ecosystems are constructed and how they function, we should be able to put them back together and make them work again. Yet while the scientists generate and refine this crucial ecological knowledge, it is the practitioners of ecological restoration who must translate and apply it. In theory, these scientists and practitioners work together to design, implement, assess, and fine-tune restoration programs. In practice, however, there is a substantial and widening gap between these scientists and their science and these practitioners and their practice.

People have been struggling to connect science to the "real" world ever since disciplinary science began, but this "science-practice gap" has become particularly problematic for applied environmentally oriented disciplines such as restoration ecology and conservation biology. This may be partly because these fields must wrestle with diverse and complex issues involving both nature and humans. Scientists and practitioners in these fields also often come from different cultures and work within different institutional settings with distinct goals, methodologies, and reward systems. Nevertheless, the urgency of today's environmental problems demands that we develop and implement effective strategies for bridging this gap so that scientists and practitioners can build more mutually beneficial relationships and accomplish more and better ecological restoration and conservation.

In this book, I explore the nature of the gap between these kinds of scientists and practitioners. How can we narrow and bridge this gap? Are the fruits of formal scientific research in these fields relevant and useful to practitioners? If and when this style of science is inadequate, are there alternative approaches that might be more effective?

Scholars from many disciplines are now analyzing and arguing over these and other scientific, philosophical, and practical questions related to the theory and practice of ecological restoration and conservation biology. While much of this literature is important and fascinating, I have found that there can be an enormous gap between how restoration and conservation are perceived and critiqued in the abstract and how they proceed in the messy real world. Similarly, there are often profound intellectual and practical differences between those who study and think about these disciplines and those who carry them out. Indeed, my personal transition from an academic research ecologist to a restoration scientist and practitioner forced me to reevaluate some of my own deeply held convictions about science, nature, and applied conservation.

Partly for these reasons, and partly because I believe it is an informative and compelling case study, I devote part 1 of this book to the story of the North Kona Dryland Forest Working Group's efforts to preserve and restore the endangered tropical dry forest within the Ka'upulehu region on the western side of the island of Hawai'i. Telling this story also enables me to offer a rare inside look at the development of, and the complex relationship between, the science of restoration ecology and the practice of ecological restoration in the context of a community-driven restoration program. Throughout this discussion and the book as a whole, I make a special effort to include the often unseen and underappreciated perspectives of practitioners—the people who design and supervise on-the-ground resource management programs as well as the ones who go home with calloused hands and muddy boots.

I begin part 2 with a more general and explicit analysis of the gap between the science and practice of applied disciplines such as ecological restoration. I then offer several strategies for more effectively bridging this gap and facilitating more positive and productive relationships between these scientists and practitioners.

I conclude with a discussion and several real-world examples of the power and promise of a more holistic, hybrid approach to restoration, which I call "intelligent tinkering" after another famous phrase written by Aldo Leopold: "To keep every cog and wheel is the first precaution of intelligent tinkering." He meant that for us to drive another seemingly unimportant species to extinction would be as foolish as disassembling a complex machine and then discarding a seemingly unimportant piece that we do not understand before attempting to put it back together again.

Unfortunately, we rarely if ever have the option of practicing intelligent tinkering in ecological restoration today because so many of the "cogs and wheels" that once made the natural world tick are now functionally or actually extinct. Leopold himself must have been keenly aware of this when he set out to restore his own badly degraded farm in Wisconsin in the 1930s. Yet even though by this time he was a highly accomplished and famous scientist, the restoration strategy he chose to employ there was much closer to intelligent tinkering in a literal sense than it was to a more formal scientific model. That is, rather than designing and implementing rigorous experiments and systematic treatments, he utilized the careful but informal, interdisciplinary, adaptive methodology typically employed by highly skilled "amateurs" undertaking such tasks as inventing or repairing a homemade gadget. The combination of Leopold's ambitious and remarkably successful on-the-ground restoration of his Wisconsin farm and his broad intellectual contributions as a scientist, educator, environmental philosopher, and writer ultimately paved the way for the subsequent development of more synthetic, applied disciplines, such as restoration ecology and conservation biology.

Leopold also believed there was a large gap between the complexity of the "land organism," as he called it, and the ability of conventional science by itself to comprehend this complexity and guide what he argued was our ethical responsibility to "doctor sick land." Over the course of his illustrious career, he thus increasingly urged both the scientific and practitioner communities to follow his lead by ignoring the "senseless barrier between science and art" and directly incorporating their personal experiences, intuitions, aesthetics, and emotions into their work. As I will attempt to illustrate throughout this book, this perspective of and approach to restoration ecology is at least as important and relevant today as it was during Leopold's time.

PART 1

Restoring Paradise

Tropical Dry Forests: Land of the Living Dead

Near the end of a long midwestern winter I got a call from a Dr. Stephen Weller at the University of California, Irvine, regarding a postdoctoral fellowship in restoration ecology at the National Tropical Botanical Garden (NTBG) on the island of Kaua'i. He explained that the original plan to fly the top three candidates to Hawai'i to interview with him and the garden staff had fallen through, but if I was still interested, the position was mine.

I had completed my PhD in biology at the University of New Mexico the previous year and subsequently landed a one-year position as a visiting assistant professor at Kenyon College, a small liberal arts school in northeastern Ohio. Things had gone well, and Kenyon had recently offered me the option of staying on to teach another year. I was thirty years old and had just fallen in love, for the first time in many years.

I looked past my teetering stacks of ungraded papers and out the dingy window of my campus apartment. It was another gray day, and the street was full of rusty cars and dirty slush. I had never been to Hawai'i and had virtually no knowledge of nor interest in our fiftieth state. (Before Kenyon offered to extend my appointment, I had frantically applied for this and dozens of other jobs in a desperate attempt to stave off the academic career–killing condition known as unemployment.) I had never met Dr. Weller, had no experience in restoration ecology or tropical biology, and, until this job came up, had never heard of the NTBG or even the island of Kaua'i.

I thought for a good three seconds about staying in Ohio another year and then accepted Steve's offer.

Three days after we were married, my wife and I flew to Kaua'i, in August 1996. Since my postdoc fellowship didn't officially start for another week, we decided to play tourist. We spent our days at the ocean, bronzing on the sand, snorkeling, and bodyboarding; we spent our nights on the lanai of our 'ohana apartment, gawking at the surreal ocean sunsets, listening to cheesy Hawaiian music, and drinking too many mai tais. One day near the end of that week, while floating in the sun-drenched ocean, I felt a weight that I hadn't even known was there suddenly lift off my neck and shoulders. I felt light and free and happy and thought that weight was gone for good.

The next month, I met with Steve Weller and two NTBG employees, Dave Lorence and Tim Flynn, to begin mapping out a restoration plan for the Garden's Ka'upulehu Dry Forest Preserve on the island of Hawai'i. Steve, a full professor of ecology and evolutionary biology at UC Irvine, was a leading expert in the study of the evolutionary genetics of plant reproductive systems. Dave, the NTBG's senior research botanist, specialized in plant systematics in general and the floristics of Pacific islands in particular. Tim was the curator of the NTBG's herbarium and an expert student of both the native and alien flora of Hawai'i.

The next day, the four of us flew from Kaua'i to the town of Kailua-Kona on the dry, leeward side of the island of Hawai'i. Looking out the window at the Kona landscape as we slowly taxied toward our gate, I saw a brown, tough-looking grass growing right through the cracks in the macadam runway. This grass was the only thing growing in the raw black lava flows surrounding the runway.

"What is that?" I asked Dave, pointing out the window.

He sighed, removed his glasses, and rubbed his forehead. "*Pennisetum setaceum*—African fountain grass," he said with a shudder, not even bothering to look out the window.

Before moving to Hawai'i, I had read about the great forests that once covered the dry lowland sections of the Big Island. These forests were the favorite place in all the islands of the great Hawaiian king Kamehameha I, the only Hawaiian to conquer all of the competing tribes and unite the entire island chain under his rule. In the early twentieth century, the famous English botanist Joseph Rock, author of the classic *Indigenous Trees of the Hawaiian Islands*, noted that there were more native tree species in these communities than anywhere else in the archipelago. Yet by the time Rock arrived in Hawai'i, most of the islands' original dry forests were long gone, and feral and domesticated herds of goats and cattle continued to ravage the few dry forest fragments that remained. These and other exotic animals also facilitated the subsequent invasion of these forests by noxious alien plants such as prickly pear cactus (*Opuntia ficus-indica*), thorny lantana shrubs (*Lantana camara*), and fast-growing Christmas berry trees (*Schinus terebinthifolius*). But although these alien species and Kona's booming tourist economy still wreak havoc on the region's remaining patches of native dry forest, today their single greatest threat is fountain grass.

If I were an alien species attempting to colonize and invade new territory, there's no place I'd rather be than the Hawaiian Islands. First, there is the islands' justly famous mild, benevolent, and stable climate. Second, there is a lot of ecological elbow room here. Because of Hawai'i's extreme geographic isolation, very few species were able to disperse to these islands and establish themselves. When the Polynesian sailors first reached Hawai'i's shores some 1,500 years ago, the only other mammals present were the Hawaiian monk seal (*Monachus schauinslandi*) and two species of hoary bat (*Lasiurus cinereus*), and there were no reptiles or amphibians at all. And because they evolved without the intense competition and predation pressures of the more crowded and diverse mainland systems, Hawai'i's native species tend to be wimps relative to their alien competitors and "ice cream" to their alien predators.

Hawai'i's native raspberries (*Rubus hawaiensis*), for instance, have essentially no thorns. Until recently there were no animals around to eat them, so natural selection favored individuals that put the energy formerly spent on making thorns into something more useful, such as making more berries. In contrast, having evolved in places with intense herbivory and competition from other plants, the exotic raspberries in Hawai'i are thorny and much more aggressive than the native species. One of these invaders, the dreaded Himalayan raspberry (*Rubus ellipticus*), has huge thorns that I quickly discovered can cut through flesh like concertina wire. Not surprisingly, this species has spread throughout the Hawaiian Islands like, well, a noxious weed.

Still another major advantage of landing in Hawai'i is the golden opportunity to escape one's troubles. (An ancient Hawaiian saying, *Lele au la, hokahoka wale iho*, translates as "I fly away, leaving disappointment behind.") For example, fountain grass is apparently a minor component of the savanna in its native region of North Africa. Its distribution and abundance are presumably held in check by Africa's many other grass species, its rich diversity of herbaceous insects and mammals, and a variety of pathogens and diseases. But when fountain grass began spreading across the leeward side of the Big Island, it apparently encountered only sunshine, open space, and Hawai'i's famous aloha spirit.

To avoid extinction, plants subjected to long periods of heavy and sustained herbivory must develop effective coping strategies. Like many other such plants, over its evolutionary history fountain grass evolved a two-tiered approach: be as unpalatable to herbivores as possible, and bounce back quickly from herbivory and other disturbances if and when they do occur. Thus, while there were (and still are) lots of cattle and goats roaming around the Big Island, clumps of mature fountain grass were about the last thing they wanted to eat. Although these animals will eat this species (especially the emerging, relatively tender new shoots) if they are hungry enough and there is nothing else around, fountain grass, like many other noxious weeds, can withstand, and even thrive under, intense levels of ungulate herbivory.

Conversely, most of Hawai'i's plants that evolved from mainland ancestors, such as the raspberry, eventually lost their ability to deter and withstand being eaten. Many of the plant species that evolved entirely within Hawai'i (i.e., those whose ancestors were preexisting Hawaiian plants) simply never developed any mammalian defenses in the first place. The net result of these contrasting evolutionary histories is that the cattle, goats, and other alien animals will search through acres of thick, raunchy fountain grass to find and devour a few delicious and nutritious leaves of some forlorn and defenseless native species.

The story of fountain grass's invasion is representative of countless other deliberately introduced exotic plants within and beyond the Hawaiian Islands. At some point in the past, somebody wanted a new plant for his or her home or business that promised to be useful, valuable, pretty, or novel. Many of us do more or less the same thing today when we search for new plants to put in and around our homes. Most of the time, these botanical adventures are perfectly harmless—the new plants either do what we wanted them to do or fail to thrive and eventually fade away. However, in a small but significant number of cases, we get much more than we bargained for.

It is not hard to imagine why someone chose to bring fountain grass to Hawai'i. When this hardy species is well watered and fertilized, it produces long, lovely clusters of bright green blades and "fountains" of spikes covered with attractive purple flowers. Even today fountain grass is often prominently featured in the manicured ornamental landscapes of Kona's expensive homes and industrial developments—the ecological equivalent of planting star thistle in Washington or kudzu in Georgia.

For many years after it was first planted on the Big Island, around the turn of the twentieth century, fountain grass apparently just sat there. Many aggressive alien species exhibit this so-called lag phase, in which they do not spread much beyond the site of their first colonization of new territory. The lag phase may last for years, decades, or in some cases even centuries. But then, for reasons that are not well understood, sometimes so slowly that no one notices, sometimes so explosively that everyone is forced to pay attention, the biological invasion begins.

Fountain grass began spreading across the leeward side of the Big Island sometime after 1915. Like many invasive plants, this species is capable of rapid and prolific reproduction and dispersal. An individual clump can produce tens of thousands of parachute-like seeds that, with the slightest breeze, are literally gone with the wind. The seeds may also be effectively dispersed by humans and our accoutrements (vehicles, equipment, clothing, etc.), water, and possibly small animals such as birds, rodents, and insects. After dispersal, fountain grass seeds may stay in a dormant but viable state for many years while they wait for favorable conditions to germinate, become established, and eventually produce thousands upon thousands more seeds. Today, fountain grass infests over 200,000 acres of arid land on the Big Island, from sea level to altitudes over 9,000 feet. Wherever it grows, this species both suppresses the establishment, growth, and regeneration of native species and greatly increases the risk of catastrophic fires.

One might assume that, given the islands' volcanic origin, Hawai'i's native species would have evolved with at least occasional fires. Indeed, for millions of years, rivers of red-hot molten lava flowed down from the active volcanoes and incinerated everything in their path until finally exploding into the sea. (This phenomenon still occurs on the Big Island, which is home to the world's most active volcano.) However, before the arrival of fountain grass and other weeds, the vegetation in many native ecosystems apparently did not produce enough fuel to carry fire much beyond the edges of the lava flows themselves. Consequently, most of Hawai'i's native species lack the adaptations necessary to withstand fires or effectively reestablish themselves in the fires' wake. Thus, unlike many mainland conservationists, who often use fire to control weeds and reestablish native species and ecosystem processes, in Hawai'i we mostly fight rather than light fires.

Kona has one of the few commercial airports left in the United States in which you still walk off the plane directly into the unfiltered outside world. Kona can get away with this because it almost never rains there, and the temperature rarely drops below 75 degrees. The second Dave, Tim, Steve, and I stepped out of our plane's air-conditioned cabin, we were engulfed by intense sunshine and hot, dry air. Walking across the runway, I was secretly disappointed that, unlike the scenes in all those old movies and television shows I had watched as a kid, there did not appear to be any beautiful hula girls waiting around to greet me with a kiss and fresh lei.

We collected our bags and rental car and headed straight for the Ka'up-ulehu Preserve. As we drove up and away from the coast, clumps of trees and shrubs began to appear within the seemingly endless fields of lava rock and fountain grass. I was also beginning to recognize both the planted and wild groves of Hawai'i's beloved (yet nonnative) food plants—bananas, breadfruit, mangoes, papaya, and coffee. (While connoisseurs argue over whether Kona coffee is truly, as advertised, the "world's best," everyone agrees that it is the world's most expensive.) My colleagues also pointed out a cosmopolitan collection of world-class weeds that had elbowed their way into the landscape. The ecologist in me couldn't help pondering how such a bizarre "ecosystem" might work: What happens when aggressive species with radically different evolutionary and ecological backgrounds are brought out to the middle of nowhere and haphazardly mixed together?

Dave turned onto a side street and stopped next to a rough young 'a'ā lava flow running parallel with the road. Most of this flow was completely barren; it looked like a frozen river of jagged black rocks. But off in the distance were a few scrubby trees growing improbably out of the lava like weeds on a gravel pile.

"'Ohe makai," Dave said, pointing to a strange-looking tree whose leaves were fluttering in the breeze.

Seeing my blank expression, Steve said, "Reynoldsia sandwicensis." I recognized this scientific binomial from my various readings, so I knew I was looking at my first native dry forest species.

"And there's a wiliwili," Tim said, pointing to an elfish-looking tree with reddish bark and no leaves.

"Erythrina sandwicensis," Steve said, rescuing me again. "The only drought-deciduous tree out here."

Having done my graduate work in the deserts of New Mexico, I was well acquainted with this clever strategy employed by many species in arid climates. Most plants acquire the carbon dioxide they need to perform photosynthesis by opening tiny pores in their leaves called stomata. However, because the concentration of water inside plants growing in dry ecosystems is usually much greater than it is outside the plant, water vapor inevitably flows out the stomata as the carbon dioxide diffuses in. As the name implies, drought-deciduous plants solve this problem by simply shedding their leaves during prolonged dry spells and thus preserving their precious water reserves. When the rains return, they quickly grow new leaves and resume photosynthesis.

"I would have thought there'd be lots of drought-deciduous species out here," I said. But everyone just shrugged, as if to say, "This is Hawai'i. Everything is different out here."

Dave pulled back onto the road, and we resumed our journey through the fountain grass and patchwork quilt of alien weeds. We climbed to about 2,000 feet and headed north toward the Ka'upulehu Preserve on the two-lane Mamalahoa Highway (Hawai'i Belt Road). After we rounded a bend in the road to the northeast, the terrain suddenly morphed from a mostly alien forested landscape to an open grassland sprinkled with scattered clumps of trees. Dave explained that even though the preserve was just two more miles down the highway, it received on average only about a tenth as much rain as the more forested region behind us. These sharp climatic gradients are common in Hawai'i and are partially responsible for the extreme physical and ecological diversity of these islands. For instance, within an area roughly the size of Connecticut, the Big Island alone contains ecosystems ranging from snow-covered mountains to soggy rain forests to bone-dry deserts.

At the next sharp turn, we pulled unceremoniously off the highway and onto some rough, gravelly lava. "There it is," Steve said, pointing out the window and upslope to our right. "The NTBG's Ka'upulehu Dry Forest Preserve."

I knew from my previous readings that this preserve lies some 2,000 feet above the coastline in the lee of the 8,271-foot summit of Hualalai Volcano. The rocks that constitute the surface substrate of this area were formed 1,500 to 3,000 years ago, when rivers of molten lava emerged from near the summit of Hualalai and then meandered all the way down into the Kona sea. One of the many meanings of the Hawaiian word *ka'u-pulehu* is "to burn breadfruit." According to this interpretation, at the beginning of a large eruption 200 years ago, the people who lived in this area prayed to Pele, the Hawaiian goddess of fire, to spare their lands and villages. She apparently heard their prayers, accepted their sacrificial offerings of burned breadfruit, and at the last minute diverted her lava around this forest. Today, a broad and barren lava field laid down during that 1801 flow forms the northern boundary of the Kaʻupulehu Preserve. So far, at least, it has protected the preserve from fires originating from that direction. Although Hualalai has since been dormant, geologists say eruptions could resume at any time.

I looked out the window and saw a sparse stand of drab, pale-green trees punctuated by largely treeless patches of both barren and fountain grass–infested lava outcroppings. After we all piled out of the car, Tim walked over and opened the "gate" by untying some twine and peeling back a crude section of loose hog-wire fence that formed the lower border of the exclosure.

I scrutinized this remnant of one of the rarest and most endangered ecosystems in the world, half expecting some kind of transcendental experience or intellectual epiphany. But what I mostly felt was hot, and what I mostly saw was a diminutive forest that reminded me of the stunted scrub oak groves that commonly grow along the East Coast of the US mainland. There were no parakeets or monkeys, no coconuts or waterfalls, no flowers or even lush foliage. After I walked through the gate and stumbled my way across the loose lava and into the forest, I found that the shade cast by these trees' sparse canopies did not provide much relief from either the intense heat or the tenacious, ankle-twisting, arm-scratching, goddamned fountain grass. I looked back across the highway and down to the turquoise sea, wondering whether there'd be time at the end of our day for a visit to one of those world-class beaches I'd noticed in the in-flight magazine on our way over from Kauaʻi that morning.

Fortunately, there was no time for me to indulge in my little daydreams because we were there to work. I was actually grateful for this because I've often found the best way to get to know new places and new people is through good, old-fashioned hard physical labor. As I soon discovered on that first day, there was always more than enough of this to go around at Kaʻupulehu, no matter how many people showed up.

Several hours later, I swung my pickax at another clump of fountain grass for the fourth and, I swore, last time. But once again the ax sliced through the fountain grass litter and glanced off the underlying sheet of lava rock, painfully twisting the handle in my already blistered hands. Throwing the ax aside in disgust, I squatted down, bear-hugged as much of the scratchy base and root crown as I could, and yanked backward. As I struggled, I saw that this clump's fibrous root network had snaked its way in, around, and through the underlying porous lava. With mounting frustration, I planted my feet even wider, bent down as low as I could, and took a long, deep breath. Using all my strength, I ripped the clump and several tightly encircled chunks of lava out of the ground and then flipped the whole wad over so it looked like an upside-down turtle. Panting, I looked at the spidery mess of torn fountain grass roots dangling from the edges of the crater I had just created. I had no idea whether those roots were capable of growing another clump of grass, but I would have bet ten to one that they could.

I stopped to drink some water, wipe the sweat and dust out of my eyes, and survey my meager progress. Like so many field projects, it had seemed like a brilliant idea during our air-conditioned, beer-driven meeting back on Kaua'i the night before. Perhaps, we had reasoned, the only way to really control fountain grass was by ripping it out by the roots. But after trying to actually implement our own plan (a mistake more experienced ecologists seldom make), I was overheated, beat up, and dubious. My plot looked as if someone had taken a pathetically underpowered rototiller and bounced and skipped his way across a few times before giving up and heading to the beach. In addition to all those taunting fountain grass roots, wisps of live grass still clung to the lava like barnacles in the many places my hands and pick could not reach. I stared at the remaining untreated sections of my plot and seriously considered the efficacy of dynamite.

Finally, I reluctantly let go of that fantasy, looked up, and studied the sea of grass surrounding the preserve. Even if we somehow miraculously managed to eradicate it from every nook and cranny within the entire six-acre exclosure, there would still be tens of thousands of viable fountain grass seeds already here in the soil, just waiting to germinate. And even if we somehow found a way to kill all those seeds, what would we do about the zillions of new seeds constantly parachuting in from the thousands of acres surrounding the preserve?

I watched my three colleagues methodically chip away at the remaining fountain grass in their plots. None of them needed to be here doing this—this kind of work was far beyond their job descriptions. Even though I was still just getting to know them, it was already obvious to me that they were exceedingly good at what they did and that their work was as much a calling as a career. If these guys—who between them had more than thirty-five years of experience working in Hawai'i—felt our little fountain grass removal experiment was worth all this blood and sweat, who was I to question it? And why was I sitting here, panting, while these relatively old geezers were still busting their butts—had too many years in academia made me that soft? I walked back to my plot, picked up my pick, and resumed my hacking.

What happened to Hawai'i's once extensive native lowland dry forests is more or less the same story of what has happened to tropical dry forests around the world. As in Hawai'i, both prehistoric and more modern cultures often chose to settle in these forests to take advantage of their favorable climates and valuable species. Because they were usually fairly open and sparse, these forests also proved relatively easy to clear and convert into grasslands or agricultural fields (or, later, golf courses and condos). Thus, even though dry forests were once the most common type of forest in the tropics, the net result of these long-standing development pressures, combined with the more recent devastating invasions by nonnative species (especially the grasses), is that today tropical dry forests are among the most endangered ecosystems in the world.

Adding insult to injury for much of what remained of Hawai'i's native dry forests, in the early part of the nineteenth century a booming market developed for the islands' sandalwood trees. Aggressive foreign traders began acquiring this species throughout the archipelago and shipping it to China. The Hawaiian chiefs, eager to acquire the military power of European firearms and the status afforded by exotic luxury items such as fine china, began paying for these goods in sandalwood. By 1827, all ablebodied Hawaiian commoners were required each year to deliver over sixty-five pounds of sandalwood to their local chief to pay off a $500,000 debt owed to American merchants by King Kamehameha II. Western observers at the time reported seeing processions of thousands of men carrying sandalwood down from the forests. By 1840, the lowlands of most of the main Hawaiian islands had largely been stripped of their once common sandalwood groves. Because performing this work was laborious, brutal, and culturally disruptive, some have even speculated that the sandalwood trade may have been partly responsible for the precipitous decline in the native Hawaiian population during this period.

Today, sandalwood is still a relatively common tree in many of the dry forest remnants on the island of Hawai'i. However, virtually all of the sandalwood trees I have seen have been small and scrubby—a far cry from the majestic specimens described and photographed by early botanists such as Joseph Rock. Since mature sandalwood trees can reproduce by sending up new shoots from their roots, I've often wondered how many of today's sandalwood trees are actually root suckers from older trees that were logged or burned long ago.

One day, I discussed this sandalwood trade while working at Ka'u-pulehu with a native Hawaiian colleague. Looking down across the miles of steep and rugged terrain that lay between us and the sea, I told him I just couldn't imagine how human beings could possibly drag or carry that much wood over such a landscape. He told me that his grandfather had told him stories of how the Hawaiian people eventually learned to pull up sandalwood seedlings wherever they saw them so that one day there wouldn't be any more trees for their children to harvest.

When Joseph Rock returned to Hawai'i in 1955, after thirty years of botanical and cultural explorations in Asia, his beloved dry forests lay largely in ruins. When he saw what had become of the forests at Pu'u Wa'awa'a in North Kona, an area he had previously called "the richest in all the territories [of Hawai'i]," he famously burst into tears. Nevertheless, I've met some old-timers who told me tantalizing tales of riding their horses as recently as the 1950s across long stretches of the leeward side of the Big Island under the shade of native dry forest trees.

Although such stories provide valuable guidance and inspiration for our present dry forest restoration efforts, given today's arid and largely treeless North Kona landscape, they can also seem like fairy tales. Even though this region still contains some of the best native dry forests left in the entire state, much of what remains consists of scattered bands of senescent trees growing within a vast sea of fountain grass. When I wander around some of the more pathetic remnants of the mighty forest that once dominated these lands, at times I feel as if I have ventured onto a hastily constructed set of some sappy western B movie, and I can almost hear the melodramatic music of its proverbial heartbreaking violin soundtrack.

After countless years of not-so-benign neglect, an unusually farsighted territorial forester named Bill Bryan realized what was happening to the last vestiges of native dry forests on the Big Island. In 1955, he successfully petitioned the Bishop Estate (the landowner now known as Kamehameha Schools) and the Board of Agriculture and Forestry to fence a relatively high quality six-acre parcel of native dry forest at Ka'upulehu and declare the area a forest preserve. Although fencing out large mammals such as cattle, pigs, and goats is now almost always the first step in preserving and restoring native ecosystems throughout Hawai'i, this was not the case in the 1950s. On the contrary, the dominant professional view during that era was that native species and ecosystems were inferior and in need of "invigoration" by stronger, more robust alien species. In fact, the territory and later state of Hawai'i used airplanes to aerially seed remote, relatively pristine native ecosystems with what today are considered some of the most noxious and intractable alien weeds. Thus, the tiny new Kaʻupulehu Preserve became one of the first areas in the entire Hawaiian archipelago to be fenced for the conservation of native species.

Bryan and his colleagues did their best to care for their new preserve by maintaining the fence and planting both native and alien trees within the exclosure. But by the time this area was fenced, fountain grass and other noxious alien plants already blanketed most of the preserve's understory. And because simply walking through fountain grass–infested areas in this rough ʻaʻā country can be extremely difficult and dangerous, it is not surprising that after Kaʻupulehu's initial fencing, little conservation work was performed there. Consequently, after a failed attempt in the 1960s to get the State of Hawaiʻi to lease the preserve, it was more or less abandoned until the NTBG took over the lease in the early 1970s. Once again, however, aside from a few botanical surveys, virtually no on-the-ground management actions were performed.

As so often happens, the impetus to finally attempt to actively restore this area came from concerned and knowledgeable local citizens. In this case, two people were primarily responsible for bringing greater attention to the perilous state of Kona's remaining native dry forest remnants in general and the Kaʻupulehu Preserve in particular: a married couple named Hannah Springer and Michael Tomich. In the late 1980s, they began educating the members of their larger surrounding community about the urgency of this situation and lobbying relevant individuals and agencies to take action before it was too late. After several years of organizing and agitating, Hannah and Michael managed to unite a broad yet diffuse coalition of people and institutions into what ultimately became the North Kona Dryland Forest Working Group. In 1995, this working group and the US Fish and Wildlife Service formally agreed to work together to preserve and restore native dry forests in North Kona.

Two fundamentally important questions for any restoration program are "Who is in charge?" and "Why are they doing this work?" The answers to these questions may range from a federal agency that is legally required to mitigate, say, the loss of wetlands due to the construction of a new highway to members of a neighborhood association voluntarily cleaning up a stretch of their local stream for purely aesthetic reasons. These different scenarios can obviously lead to different ways of designing and implementing restoration programs, as well as different roles and responsibilities for the scientists and practitioners working in these programs. Over time, as some of the original members inevitably leave and are replaced by new people with different priorities and perspectives, the unifying mission and cohesion of the founding group can splinter into sometimes divisive and contentious factions.

Community-driven restoration projects may be especially vulnerable to this kind of Balkanization because they are typically composed of a large number of individuals and agencies with their own often diverse and shifting visions, commitments, and resources. Indeed, the North Kona Dryland Forest Working Group eventually included over forty individuals representing more than twenty-five agencies and numerous individuals from very different segments of the local community. It became obvious to me shortly after I joined this group, in 1996, that significant tensions and conflicts had already developed among some of its members and their respective institutions.

As we collectively struggled over the succeeding years to hold this coalition together, I became increasingly interested in learning more about how and why this working group got started in the first place. Thus, after I had come to know Hannah and Michael through several years of dry forest meetings and collaborative projects, I jumped at the chance to sit down and discuss these issues with them one summer day in 2004 at their beautiful plantation-style home. (Their house, just two miles down the road from the Ka'upulehu Preserve, served as the location for most of the group's official meetings.) I began by asking them to tell me the story of their personal connection to Hawaiian dry forests as well as the steps that led to the creation of the North Kona Dryland Forest Working Group.

Hannah closed her eyes momentarily, lost in thought, and then began with a brief account of her family history. "Our children are the sixth generation to live here at Kukuiohiwai, our home inside the *ahupua'a* [a Hawaiian land division system somewhat similar to our modern concept of ecological watersheds] of Ka'upulehu, so our family has been residing here as a landowner since the nineteen-teens. Some of my Hawaiian lineage traces its roots to the adjacent *ahupua'a* of Kuki'o, surely at the time Captain Cook arrived, and I could give you a recitation by name of those individuals, which would be a very Hawaiian practice to indicate this is who my people are; this is why I have not only the confidence, but the accountability, to speak to you—I could go back to my seven-times great-grandparent who lived a subsistence, then later surplus, lifestyle here.

"When I was very young my family owned the Ka'upulehu lease, so one of my earliest memories is going into the forest and being surrounded by a 'murder of crows' just harassing and squawking at us, and as a child I was on the floor of the car, going, 'What are these things?' Little did I know how rare an experience that was. At the time, with the people I was with, those were just the crows—that's what they do. [Today the 'alalā, or Hawaiian crow, is extinct in the wild, and only a handful exist in captivity.]

"My mom's favorite flowers included the uhiuhi [Caesalpinia kavaiensis, a now federally endangered leguminous tree that produces beautiful red flowers and extremely hard wood that the Hawaiians used for spears, posts, and various tools and fishing implements] and the koki'o [Kokia drynarioides, another endangered tree in the Mallow family with maplelike leaves and stunning, scarlet-colored flowers; the Hawaiians used to strip off and macerate its bark to make a dye for their fishing nets], so also when I was very young, and there were more than a handful of those species out along the roadside, we'd go along the road and she'd point out particular favorites, so again, just as a child, cruising with my family, I had a wonderful insight into the members of a community that were literally fading before my eyes. Before I had any sort of academic appreciation of what was going on, I associated those flowers with my mom, so it gave me a pretty deep level of intimacy with this landscape.

"When my great-great-grandfather had the Ka'upulehu lease, one of the things that he was required to do was 'goat control.' We can look at the Kona Historical Society's records around that time, the 1910s, and see that there were 13,000 to 17,000 goat hides a year being harvested off this landscape. Part of the irony is that I stand before you now as the descendant of Hawaiians who raised goats out here. Captain Vancouver came and said, 'Hey, this is good stuff! You guys gotta raise this, 'cause us guys is coming here and gonna be wanting them!' And so our family sensibility has evolved from a time when there were no goats to a time when we said, 'Hey, let's raise some goats for y'all,' to a time when there were too many goats, let's reduce their numbers by hunting, to a place that we're sitting in now, where I certainly am among those who would champion a goat-free environment. This is sort of a metaphor for our lives in general: as we weave our way through the shifting sands and moving tides, both individually and as a community, we collectively need to change our wisdom and values.

"So as a kid I was aware of the crows and some of the pretty flowers, but the rest of the landscape was just a gray blur to me as I went about playing in my playground with the other kids. But then in the 1970s I started taking hula and I learned that an uncarved block of lama represented Laka [the patron of hula], and so I began piecing together things of my childhood with this new study I was embarking upon through hula." Lama, Diospyros sandwicensis, is a member of the Ebony family. This is the dominant native dry forest canopy tree at Ka'upulehu and throughout much of North Kona; it produces edible persimmons and very hard wood, which the Hawaiians fashioned into rafters and traps for deep-ocean fish. They also pulverized the wood and mixed it with other materials to make compresses for the treatment of skin sores.

Hannah told me that around that time she met Michael, and the two of them married and moved to the town of Volcano, on the other side of the island. Shortly thereafter, she met Lani Stemmerman, a famous Hawaiian ecologist, teacher, and environmental activist, when they both reached for the last beer in Volcano's general store. Hannah told Lani that she'd let her have the beer if she would come to Kona and teach her about the region's native plants.

"Lani just opened my eyes to the things we don't often see," Hannah recalled, "and few of us know about this landscape, and it was a wonderful complement to my family's collection of Hawaiian things from the ancient times that I had always held in high regard for their cultural value. Lani helped me see those plants as a scientist would but also as cultural resources. She helped me better understand their role and place in this ecosystem; prior to that they had just been names in books or implements on our shelves."

I asked Hannah how prominent fountain grass was when she was a kid and what her family's relationship was to this species. "When fountain grass was introduced here, whether it was when another family lived at this home, or whether it was by my great-great-grandfather's family, I don't know, but I have no doubt that it came here as an ornamental. As the story goes, the local people recognized the potentially noxious properties of the grass, so they took it out to the lava channel there in Ka'upulehu and burned it. Now I can just imagine all those little seeds going, 'All right, thanks, gang!' But that was also during the time when they were harvesting goats, so the goats were at high numbers while the fountain grass was at low numbers. Then later there was a series of goat drives to suppress the numbers of goats on the public lands—those drives are described in Michael's dad's book [*Mammals in Hawai'i*, the definitive work on this subject], so it's also true that as the goat numbers were going down, the fountain grass numbers were going up.

"We can look at the pictures in Rock's book [*The Indigenous Trees of the Hawaiian Islands*] and look into the background, look through the trees and see how there were tens of thousands of goats out there and no fountain grass. You can also look at the pictures my parents took around the time I was born [1952] and see that these are not the fountain grass lands as we see them today. My folks used to tell the story of a boat running aground at Kiholo [a small bay about ten miles away as the *'alalā* used to fly] and the person being able to follow the old government road during a bright starlit or moonlit night all the way to our house. In the mid-1980s, we went out to look for that Kiholo-Huehue trail and it took us a full day just to find it! Even though Michael and Lani are both excellent field people, it still took us three days of concerted effort to walk that trail in its entirety, which shows to what extent fountain grass has just exploded across and buried much of this landscape.

"As a descendant of ranchers, whether our ancestors introduced species such as fountain grass or some of the other now noxious species . . . that's part of our family history. But over the course of time, some of us apply our sensibilities that are acquired through the generations of gaining our livelihood from the land to improving the land in our turn, as our ancestors thought that they were doing in their turn. We might make some adjustments due to external influences like the goats, or internal influences as we evolve and see the causes and effects of our presence on the land. I think we can look to the Hawaiian culture and the *kapu* [taboo] system and its complexities and its rigor as an example of arriving in an island environment where at first food was just falling off the land with the birds and from the sea with the shellfish. But after populations were reduced over time, they responded by eventually creating a system of laws and orders that attempted to allow for sustained yields off this island ecosystem. This story is not unique to Hawai'i—wherever humans go, these impacts occur, and then the question is, 'How will we resolve them?' Michael's comment about much of the story of the Old Testament is, 'Hey, this is just an account of bad land management!'"

I asked Hannah and Michael how their growing interest in and knowledge of their local remnant dry forests led to their efforts to actively preserve and restore these lands. "One of the places that we enjoyed going with Lani was the NTBG plot; she knew more about that exclosure than we did," Hannah replied. "We started doing little service projects, weeding around our favorite specimens, doing slide shows, Michael as a fireman, me as a public speaker, bringing people's attention to the dryland forest."

"We also started talking with the landowners, the federal agency people, and the land developers about the importance of conserving these remnant native dry forests," Michael added, "and in 1989 I wrote up some fire management recommendations. But back at that time, the conservation community thought the dryland forest situation was hopeless—there were some very seasoned people who came out here and looked at these forests and the extent of the fountain grass invasion and basically said, 'Forget it!' So while some people knew what was happening out here, and how much was being lost, the general consensus at that time was that there was nothing that could be done about it; it was a lost cause, or at least it wasn't worth the effort to try to restore it."

Hannah laughed and continued the story. "Around that time, I met Roger Harris when he was working on a big resort development down on the coast, and he asked me to do a slide show on west-side botany and cultural geography. Then, later, he became the land manager for Potomac Investment Associates, which has the sublease on the lands just below the highway adjacent to the NTBG plot, and he asked us to write up our perspectives on the cultural geography of that region and invited us to go into the dryland forest there and 'putter.' In 1994, PIA erected a small fenced exclosure in their land out there, and it was such a pleasure to work with Roger because he understood why we needed to protect these remnant dryland forests from invasive grasses, and how those grasses are such a boon to wildfires. Before they did any planting in that exclosure, they spent about a year weeding it by hand—they were practically out there with a magnifying glass, dental tools, and tweezers! The level and intensity of the work they did out there is unrivaled by anything else I have ever seen.

"So then one day I was giving another slide show at a luncheon that included people from Kamehameha Schools, and I was talking about the irony for me as a Hawaiian that the *haole* land developer, who I'm not supposed to trust and who I'm supposed to resent on my landscape, was giving me the opportunity to learn more about and attempt to restore a native dry forest, and I think the Kamehameha Schools people were saying, 'Well, maybe we should start doing something too!' [Kamehameha Schools was explicitly established to create educational opportunities for and improve the well-being of the native Hawaiian people.]

"So their interest was piqued, and there was this confluence of Michael and I being on this learning curve with Lani, and it became politically ripe for Kamehameha Schools to take an interest in our dry forest work, and the federal agencies were starting to see what was going on here. Then we got that first infusion of money through the US Fish and Wildlife Service to explore the feasibility of preserving and restoring native dry forests. So we hired a facilitator and sent out invitations to various stakeholders to work with us, including some of the landowners who we recognized as having good dryland forest remnants on their holdings, various state and federal agencies, The Nature Conservancy, and Kamehameha Schools.

"We built a matrix that compared the characteristics of different remnant dryland forests in this region—things like their diversity, proportion of native species, natural reproduction, extent of fountain grass coverage, and, most important, landowner willingness to engage! [Much as on the southwestern US mainland, many of the region's ranchers, hunters, and local citizens are at best suspicious of the federal government in general and "environmentalists" in particular.] It was through this process that we ultimately arrived at the NTBG plot at Kaʻupulehu.

"We also identified another remnant forest within a *kīpuka* not far from the NTBG plot, a storied, remarkable place, according to those of Hawaiʻi who were here before me. It had *kokiʻo, uhiuhi, ʻaiea* [*Nothocestrum breviflorum*], *kauila* [*Colubrina oppositifolia*]; everybody was out there! The Kaʻupulehu lava flow formed what we thought was a natural firebreak around it, but then in May of 1993 fire broke out inside the *kīpuka*, and when it was done, the best of the last of those trees was gone. We have every reason to believe it was a deliberately set fire, but no one ever came forward and no suspects were ever identified. Shortly after that fire, all those trees finally became listed as a federally endangered species.

"We consciously decided to have those first Fish and Wildlife–sponsored meetings here at our home, rather than at an agency office. People come with different demeanors into a home than they do into a conference room—we would say, 'Check your boots, slippers, guns, and egos at the door and come inside!' We also deliberately structured the meetings as roundtable discussions so we could get the agency guys, the science guys, the conservation guys, the cowboys . . . all the players talking to one another."

When I asked whether it had been difficult to gain the initial cooperation of the ranchers to work with their conservation-oriented group, Michael recalled that "the ranchers weren't adversarial, but they might have been a little bit threatened and afraid of losing some grazing land. Plus they had an attitude of 'Show me what you guys can do before I give up any land—show me you guys can handle this.' But actually Hualalai Ranch played a big part in the beginning by donating water."

"From the beginning of this project [the mid-1980s] right up to the present [2004]," Hannah continued, "one of the areas of tension in discussion with the cowboys is that some of the folks with tremendous academic backgrounds have little to no actual land management experience, and consequently don't always appreciate how difficult it can be to put academic theories into practice! Cowboys might not have much academic background, but they have a lot of experience working the land. I know one of the things that used to bug them was when the enviros would say things like, 'Well, we'll let you run your cows here' [for example, to help create and maintain firebreaks]. And they're going, 'It's really steep, rough country—you're not doing us any favors; our cattle don't want to be there. You should pay us to graze our cattle there!' So maybe there wasn't enough give and take; maybe it was take and take. But what was really important was that at least they were at the table and dialoguing with us about the potential conservation of native dry forests."

In 1995, Hannah and Michael's newly formed North Kona Dryland Forest Working Group decided to begin by carefully documenting and assessing the current situation in the NTBG plot at Ka'upulehu. Dave Lorence and Tim Flynn flew in from Kaua'i and performed detailed vegetation surveys inside the exclosure and, for comparative purposes, in the adjacent land outside the exclosure. The good news, they found, was that this preserve still contained an impressive diversity and abundance of native dry forest plants, including four federally endangered canopy tree species. The bad news was that it had essentially become a forest of the living dead.

While the fence had effectively excluded large animals such as cattle and goats, it had of course done nothing to stop fountain grass and other noxious weeds, rats, mongooses, and swarms of exotic birds, insects, and who knows what else from entering and wreaking their typical ecological havoc. Dave and Tim also found that the native canopy trees had long since failed to regenerate—there were virtually no seedlings or saplings of these species in the entire exclosure. Their comparisons of the present flora with past surveys of the preserve further revealed that some native trees considered common only twenty years earlier had been almost completely extirpated, and almost all of the native trees planted by Bill Bryan and his colleagues were dead. While it was a miracle that so many native understory species were present at all, these plants were largely confined to some scattered lava outcrops that would almost certainly be smothered one day by the ever-encroaching swaths of fountain grass. Finally, there was the very real possibility that one day the lucky streak would end and the entire forest would go up in smoke.

After much debate, the working group agreed to begin its restoration of this preserve by waging war on fountain grass. After some trial-and-error experimentation, they decided to attack this species in a two-step process. First, they would use weed whackers to cut the grass as close to ground level as possible, to remove the rank layers of mostly dead vegetation and expose each clump's inner core. Second, once the grass began flushing back, they would spray this new and relatively vulnerable growth with herbicides, which they hoped would kill the grass for good.

To implement this plan, in the winter of 1995 the working group hired two local guys to cut and spray the fountain grass within the entire six-acre exclosure. Thus, after seemingly endless rounds of talks, meetings, hand-wringing, arm-waving, finger-pointing, surveys, reports, site visits, VIP tours, and speeches, forty years after the territorial forester Bill Bryan first fenced this remnant forest, the actual restoration of the Ka'upulehu Preserve had finally begun.

By the time of my first visit, in the fall of 1996, the lower, most accessible sections were largely free of fountain grass. But as we walked away from the highway and into the preserve's deeper reaches, we saw more and more patches of living fountain grass, and in the most remote and inaccessible areas there were sections that looked as if they had never been cut or sprayed at all. Many of these apparently untreated areas contained jagged and steep sections of 'a'ā that would have been treacherous to navigate even in the absence of fountain grass; attempting to traverse them and their blankets of lava-obscuring grass required heroic determination, a masochistic personality, or both. But I hadn't realized how relatively easy it was to merely walk through fountain grass–infested lava until I spent my first day wielding a hot and clunky weed whacker and lugging a heavy backpack tank full of herbicide.

I never got to meet those two guys (after that job, they apparently never expressed any interest in working at Ka'upulehu again), but I often wondered whether they knew anything about the significance of their work or were simply trying to make a buck off the "crazy haoles." Yet the more time I spent in the trenches of the fountain grass war at Ka'upulehu, the more I realized that regardless of their motivation and state of mind, whatever those guys were paid, it wasn't enough.

In the end, it took them over ten months to "finish" cutting and spraying all that fountain grass. The total bill came to more than $5,000 per acre. Some felt this was a ridiculous waste of taxpayer money, while others saw it as a bargain compared with the millions of dollars the US Fish and Wildlife Service often spends in Hawai'i and elsewhere on arguably misguided efforts to conserve a single charismatic yet probably doomed endangered species. Even within the Hawaiian conservation community, there were and are people who think it foolish to work in these "basket-case" systems and who believe we should spend our precious conservation dollars on the more pristine, relatively intact, upland ecosystems. On good days, I felt this battle plan would be a tragic mistake; on bad days, I was almost willing to argue for it myself.

As I walked around the forest at the end of my first week there, I couldn't help wondering whether fountain grass would soon reinvade the entire Ka'upulehu Preserve (including our hand-cleared plots) and if all that hard work would in fact wind up being a noble but foolish effort. But over time I began to realize that regardless of its ultimate biological effectiveness, this war on fountain grass had already produced some important results. First and foremost, as I was to repeatedly discover through many years of North Kona Dryland Forest Working Group meetings, in contrast to those endless hours of what could often devolve into emotionally charged and divisive debates, getting out there and doing something together was both unifying and inspiring. And second, because one could now walk through large sections of the preserve without risking life or limb, more and more people were able to come out and see this forest for themselves. Consequently, the working group grew larger and stronger, and its members became increasingly energized and hungry for more concrete, on-the-ground action.

Not long after our little manual fountain grass removal experiment, the working group decided the time had finally come to stop killing and start planting. As luck would have it, Dave and Tim had already collected many native tree seeds from the preserve and brought them back to the NTBG nursery on Kaua'i for propagation. When the group learned that there were now more than 400 potted plants from that seed collection, everyone agreed that the next major step should be to bring those plants home and plant them back in their native soil. We all also agreed that we should design and implement this outplanting project by utilizing and applying the best scientific knowledge and practices. As the newest member of the team and the slowest to say no, I suddenly found myself in charge of the whole operation.

Chapter 2

Let's See Action! Planning and Implementing a Research and Restoration Program

I didn't realize how different the islands of Kaua'i and Hawai'i were until I flew back to Kaua'i after that first trip to the Ka'upulehu Dry Forest Preserve. While both have substantial ecological and cultural intra-island variation, as a whole they are literally, and in many ways figuratively, at opposite ends of the chain of major Hawaiian islands. For example, with its vast unsettled forests (if you don't count the scattered bands of off-the-grid squatters living on mangoes and marijuana), extensive molten red and jagged black lava, and local politicians openly packing illegal firearms while trolling for votes along the highway, much of the Big Island has a frontier, Wild West feel. Conversely, most of the inhabited, accessible parts of Kaua'i (the "Garden Isle") are relatively soft and lush and are overrun with highly domesticated, fanny-pack-toting tourists.

These two islands are only a small part of a long chain of volcanoes that begins near the Big Island and runs in a northwesterly direction for nearly 4,000 miles. Hundreds of miles beneath the Big Island's active volcanoes lies a stationary "hot spot" that continuously pushes magma up through the Pacific Plate. This tectonic plate, which drifts in a northwesterly direction at an annual rate of about three and a half inches, has, over millions of years, rafted away each of the new volcanoes that have formed over the hot spot. Thus, as one moves northwest up the chain, each volcano (an "island" while it lies above the ocean and a "seamount" when it sinks below) is progressively older and more weathered. The oldest known seamount, near the northern end of the chain not far from the Aleutian Islands, is about 80 million years old, while the actively erupting Lō'ihi Seamount, eighteen miles off the Big Island's southeastern coast and 3,200 feet below

35

sea level, is predicted to emerge as the newest Hawaiian island within the next 200,000 years.

This geologic history helps explain why both the physical environments and the ecological communities of the Big Island and Kaua'i are so different. Because Kaua'i (about 5 million years old) has been above sea level substantially longer than the Big Island (about 400,000 years old), there has obviously been much more time for processes such as erosion and island subsidence (caused by the massive weight of the volcanoes pressing down on Earth's underlying crust) to shape its landscape. In contrast to the Big Island's towering pair of young 13,700-foot volcanoes, the highest elevation on Kaua'i is now less than 5,250 feet. Instead of raw, unweathered lava fields and wide, sloping sides, Kaua'i has heavily eroded mountains with knife-edge ridges, deep-soiled valleys (including the magnificent Waimea Canyon—aptly dubbed the "Grand Canyon of the Pacific"), and broad, world-class beaches.

Long before what would become the Big Island even began bubbling up from the sea, living species were colonizing and evolving on Kaua'i. This is a major reason why the biological richness and endemism within a given ecosystem on Kaua'i tends to be much greater than it is on the Big Island. In fact, many of the older native species that exist today on the younger islands have been traced back with molecular techniques to ancestral species that originally lived either on Kaua'i or on the now tiny, severely eroded Necker Island to its northwest. In other words, in these cases, descendants of the original continental colonists on the older Hawaiian islands hopscotched their way across the interisland ocean channels and eventually colonized the younger islands after the islands formed over the hot spot and began their slow northwestern drift.

We can also estimate the time at which a new species first reached Hawai'i by analyzing the accumulated differences in DNA between the extant island species and the presumed ancestral continental species. Similarly, we can compare the DNA sequences of the same or closely related species now living on different Hawaiian islands to unravel the sequence of "founder events" that ultimately led to their establishment on the different islands. For example, comparative molecular analyses indicate that the original founders of what would ultimately evolve into Hawai'i's famous alliance of silversword plants arrived at least 6 million years ago, perhaps on an island older than Kaua'i, such as Necker.

These data further suggest that the subsequent colonization and diversification of these bizarre yet beautiful species on the other major Hawaiian islands were brought about by thirteen separate interisland dispersal events. Today, there are twenty-eight different silversword alliance species in the Hawaiian Islands, and all but two of the thirteen species on Kaua'i are single-island endemics (i.e., not only are they found only on these islands, but also within this archipelago they exist only on Kaua'i). Yet, despite the Big Island's much larger size and ecological diversity, apparently there has been enough time for only seven silversword alliance species to disperse to or evolve on it, and four of these seven are found on at least one of the other Hawaiian islands.

A few days after we returned to Kaua'i, Steve Weller flew back to California, Dave Lorence and Tim Flynn went back to their respective desks, and I returned to unpacking and reading the newly arrived scientific books and papers on restoration ecology and Hawaiian conservation biology I had shipped to myself just before leaving the mainland. Now that I was about to embark on my first real restoration project, I was eager to learn as much as I could from this literature as well as look for particular areas in which I might be able to build on and add to our existing scientific knowledge in these disciplines.

The more I learned about Hawai'i's endlessly fascinating evolutionary and ecological history and present conservation crises, the more I came to appreciate the larger importance of past and present efforts to preserve and restore that remnant native dry forest at Ka'upulehu. Yet, as engrossing as it all was, I knew that I also needed to think through the nuts and bolts of my postdoctoral fellowship at the National Tropical Botanical Garden.

In 1964, the United States Congress officially chartered the Pacific Tropical Botanical Garden as a privately funded research and education institution. Today this organization, which is the only major US botanical garden located in the tropics, includes four gardens and three preserves in Hawai'i and one garden in Florida. (After the organization acquired The Kampong in South Florida, Congress changed its name to the National Tropical Botanical Garden.) As detailed in its original congressional charter, the NTBG's official purposes included the establishment and operation of educational and scientific centers related to tropical botany, as well as a mandate to "collect and cultivate tropical flora and to preserve for the people of the United States species of tropical plant life threatened with extinction."

In an effort to fulfill these goals, over the years the NTBG has created numerous research, education, and outreach programs that have extended well beyond the borders of the 1,800 acres contained within its own gardens and preserves. The NTBG was particularly successful at finding and propagating dozens of new Hawaiian species (as well as dozens of species previously presumed extinct), and the often heroic, cliff-dangling efforts of its plant collectors were widely featured in documentaries and popular magazines. Nevertheless, in the mid-1990s, Dr. William Klein, the NTBG's president and executive director, felt that the organization should be doing more to help preserve Hawai'i's increasingly endangered flora. He eventually hired Steve Weller to serve as the NTBG's first McBryde Chair in Hawaiian Plant Sciences. However, given Steve's mainland base, his existing research program in the evolutionary genetics of plant breeding systems in Hawai'i, and his myriad responsibilities as a full professor at the University of California, Irvine, they decided the best way to engage the NTBG in more on-the-ground Hawaiian conservation efforts would be to hire a Hawai'i-based McBryde Postdoctoral Research Fellow in Restoration Ecology.

To be successful in the highly competitive world of modern academia, scientists usually must acquire additional teaching and research experience after completing their PhDs. The most common pathway for accomplishing this is to teach your brains out for a year or two as a visiting assistant professor (as I had done at Kenyon College) and then research your brains out for a year or two as a postdoctoral research fellow. The teaching part of this equation is relatively straightforward—some college hires you to teach and, ready or not, you get up there and do it.

However, there are many different postdoctoral research models and pathways. Some postdocs essentially apprentice with their mentors in order to learn specific technical skills or improve their ability to write competitive grant proposals, design and perform research, and publish in academic journals. In other cases, the relationship is much looser, and the postdocs are largely free to pursue their interests in a partially or completely autonomous fashion.

Even before he hired me, Steve made it clear that while he had several potential projects in Hawai'i for me, and he would always be happy to advise and assist me in any way he could, I would also be free to tailor the specifics of my postdoc in whatever manner I felt best matched my strengths and interests. Although I found Steve's research fascinating, I just didn't seem to have the necessary interest and aptitude to plunge into study of the evolutionary genetics of plant breeding systems. Yet at the same time, I wasn't at all sure what kind of research program I could or should carve out of this strange new world of ecological restoration in Hawai'i.

I stared out the window of my new office in the NTBG's administrative headquarters on Kaua'i's sunny south shore, feeling very much alone. Below me lay a vast, beautiful, but empty ocean. It suddenly dawned on me that while Tim liked to boast that he was the best herbarium curator for thousands of miles south of the botanical garden (the nearest inhabited island in that direction is more than 2,000 miles away), he could actually broaden his claim to include all of Kaua'i, and I could likewise claim to be the best postdoctoral restoration ecologist in the entire Hawaiian archipelago.

Ironically, however, I knew that the two Lāwa'i Valley gardens that lay between me and the ocean were successful examples of what we might now consider a form of ecological restoration. The first one, the 250-acre McBryde Garden, was named for the descendants of the family that once cultivated sugarcane throughout this entire valley. Since the NTBG acquired this parcel shortly after its 1964 congressional charter, its staff has steadily converted that former cane field into a Noah's ark for tropical plants, which now includes an ethnobotanical Canoe Garden, featuring plants that the Polynesians brought to Hawai'i in their voyaging canoes; systematically important plant collections; and the world's largest living collection of native Hawaiian plants that includes endangered and extinct-in-the-wild specimens.

Between the McBryde Garden and the sea lies the 100-acre Allerton Garden. Frequently cited as a "masterpiece of garden paradise and tropical romanticism," it was painstakingly conceived and built over a forty-year period in the mid-twentieth century by Robert Allerton, a Chicago philanthropist, world traveler, and art lover, and his adopted son, John, a skilled landscape architect. The Allertons and their Hawaiian helpers ultimately crafted a "jungle" full of exquisite collections of tropical plants from around the world. They also intertwined these plantings with European and Asian statuary, reflective pools, and various gravity-fed water features strategically situated to drown out the sound of the trucks hauling sugarcane out of the surrounding fields.

The first time I wandered through some of the Allerton's meandering paths, I immediately felt as if I had been transported to another time and place. This garden is in fact so convincing and encapsulating that many prominent filmmakers have used it as a backdrop. Perhaps most famously, the scene in *Jurassic Park* where the heroes find the dinosaur eggs was shot among the huge buttressed roots of the Allerton's Moreton Bay fig trees. *Fantasy Island*'s "De plane! De plane!" opening was filmed in a field adjacent to these trees.

When I encountered groups of tourists on one of the NTBG's guided walking tours of the Allerton Garden, I sometimes overheard them telling each other that the garden had evoked some highly personal, deeply moving emotional or spiritual sensation. The *Jurassic Park* fig trees seemed to be a particularly popular place for this to occur. Over the years, I witnessed several people have such experiences there and later heard them wax poetic about such things as how they had finally connected with the primeval tropics, spiritually bonded with the real Hawai'i, and so forth. I never had the heart to tell them that those "ancient" fig trees had been brought over from Australia and planted about seventy years ago, and that nothing within that garden was actually Hawaiian. However, after a few largely futile attempts, it didn't take me long to realize that compared with the splendor and charms of the Allerton Garden, getting tourists, or even the local public, interested in the McBryde Garden's scruffy native plant collection and ecological restoration in general was a tough sell.

I tried once again to snap out of my window-gazing daydream and focus on the tasks demanding my full and immediate attention. First, I needed to plan and implement next month's Ka'upulehu outplanting of all those potted plants down in the nursery, which, the plant propagator had not so subtly informed me, had been clogging up precious bench space for over a year and were now starting to die. But before I could tackle that job, I needed to round up and study all the previous reports and data from Ka'upulehu in order to at least semi-intelligently begin designing a broader research and restoration program for that entire preserve.

Second was an equally pressing task that involved more fieldwork, data analysis, and experimental design for a mesic forest on Kaua'i within a drainage called Mahanaloa Gulch. Before he flew back to California, Steve had taken me to a remote valley on the northwestern side of the island and shown me around a relatively intact remnant forest clinging to that gulch's steep north-facing slope. To me, Mahanaloa was a beautiful but bewildering quilt of species assemblages—just when I thought I was finally beginning to at least recognize the dominant native tree species, I'd walk another hundred yards down the valley and suddenly find myself in an entirely different forest! Even more disconcerting were the several occasions when Steve pointed out species I had just seen at Ka'upulehu but, because they looked so different to me, and in some cases had taken on completely different growth forms (e.g., what had been a small shrub in the dry forest had morphed into a vinelike tree here), I never would have recognized on my own. This was my first direct, personal experience of how profoundly evolution could shape the morphology and ecology of species as they dispersed across and evolved on the different Hawaiian islands.

What also struck me about Mahanaloa's flora was that every other species seemed to be another federally endangered single-island endemic. Although we hadn't explicitly thought it through at that time, we later deduced that since the Hawaiian flora, which has the highest proportion of endemism and endangerment in the world, reaches its greatest level of endemism and endangerment on Kaua'i, and that on this island the flora reaches its greatest levels of these variables in Mahanaloa Gulch, to the best of our knowledge that remnant forest contained the most endemic and endangered flora in the entire world. Tragically, aside from a small and neglected state exclosure, the rest of Mahanaloa's forest was unprotected from the ravages of Hawai'i's ubiquitous exotic ungulates, which in this case included apparently abundant populations of feral pigs and deer, given the extensive herbivory, game trails, and droppings we saw.

Prior to hiring me, Steve and Dave had written a major National Science Foundation grant application in which they proposed to perform parallel ungulate exclusion experiments at Ka'upulehu and Mahanaloa Gulch. Although NSF had not funded their grant (few such proposals are funded on the first round), they had kindly invited me to be a coauthor on a revised version of this grant that was due in a few weeks. We believed this research was important for both the science and practice of restoration ecology and the conservation of Hawai'i's endangered flora. However, as is the case with so many of these kinds of projects, we knew that without the kind of substantial external funding and credibility that NSF grants provide, the probability that our proposed research would ever happen was close to zero.

To address the concerns of the panel of expert reviewers who had rejected the first submission, we needed to include more preliminary data on the ecological effects of that state ungulate exclosure as well as a more detailed and focused overall research plan. Thus, I needed to go quickly back to Mahanaloa Gulch with Dave and Tim to collect those data so we would have sufficient time to analyze and incorporate them into our revised NSF grant application.

Given that my postdoctoral fellowship was funded for only a year or two, depending on how well things worked out, I also needed to quickly decide whether and to what extent to get involved in several smaller, more local restoration projects. Steve had helped design one of these projects, which involved an experimental effort to establish a demonstration native forest community within an established grove of aggressive alien species in a remote section of Lāwa'i Valley. While this and some of the other projects were interesting and seemed to have great public education and outreach potential, I wasn't sure whether these kinds of "gardening adventures" (as some of my harder-nosed colleagues patronizingly called them) would have much scientific or conservation value.

Last on my to-do list were several technical manuscripts from my PhD research. Before I left, my graduate school advisor had explained that it becomes increasingly difficult to publish one's dissertation research after graduation (you are too busy doing new things, you lose the necessary focus and technical expertise, your results become dated and passé, etc.), yet the more publications you have, the more competitive you are to prospective employers. She also told me that the best time for an academic to crank out publications is as a postdoc, as it's the only time when you don't have to teach, advise students, and serve on committees.

Despite my best intentions, as she predicted, I did not make much progress on my dissertation papers during my year at Kenyon College because I was far too busy teaching. However, I had managed to submit a few papers during the previous summer, which had since come back with commentary ranging from "We would be happy to publish this pending some relatively minor revisions" to "We don't want to see this manuscript in any form ever again!" I wanted to at least tackle the papers in the former category before my window for resubmission closed. Even though that work now seemed as if it had been performed by another person in a different world, and in truth I was no longer very interested in the demography and genetic structure of desert soil seed banks, I was nevertheless both selfishly (for the sake of my own career) and ethically (I felt a moral obligation to disseminate the results of all that publicly funded science) determined to see that research through to publication. I also knew from past experience that what might seem like "relatively minor revisions" to the journal reviewers and editors would probably require several long days of uninterrupted work on my part.

I finished making my to-do list, feeling simultaneously relieved and overwhelmed to have it all spelled out in front of me, and then looked back out the window at the shimmering ocean. Part of me yearned to just bolt out the door, grab my wife, and race down to the exquisite private beach where the Allerton Garden ran into Lāwa'i Bay. Then, of course, we still had our long personal to-do list with its own pressing items, such as finding a place to live and buying a car (nontrivial tasks on Kaua'i on a postdoc salary). While Dr. Klein had kindly put us up in the 'ohana apartment attached to his house and temporarily lent us one of the NTBG's old beater cars, the last thing we wanted to do was infringe upon his hospitality.

I finally decided that since I had been working nonstop for the past few weeks, once I installed my new statistics software so that I could hit the ground running tomorrow morning, it might be okay to sneak in a swim before sunset. In addition to needing this software to redo some of the statistics in my dissertation papers, I had promised Dr. Klein that I would look at some of the NTBG's data sets that he felt could benefit from some "more formal and rigorous analyses."

While I did have a decent background in and working knowledge of the basic quantitative principles and statistical procedures that serve as the theoretical foundations for much of the disciplines of modern academic ecology and evolution, I was not particularly good at or even interested in those kinds of analyses. On the contrary, I was much more of an empirical "get out there and get dirty" kind of ecologist. In fact, most of the results and conclusions of my dissertation research on soil seed banks came from a largely brute-force effort in which I meticulously processed thousands of samples of soil that I had laboriously collected from and backpacked out of a remote and relatively pristine desert field site.

My motivation for this research stemmed in part from the fact that much of what we knew about the ecological and evolutionary dynamics of seed banks came from sophisticated yet abstract models of how these seed populations might form and evolve over space and time. While the complex mathematics underlying these elegant models was often over my head, I was intuitively skeptical of any purely theoretical attempt to reduce the vast complexities of nature to a series of abstract equations. Indeed, my empirical data ultimately contradicted some of the most important insights and conclusions generated by much of the relevant theoretical seed bank literature.

This kind of discrepancy between the results and implications of theoretical versus empirical science frequently occurs in the disciplines of academic ecology, restoration ecology, and conservation biology. Not surprisingly, in such instances the theoreticians tend to assume that the empirical data must be flawed, and the empiricists tend to assume that the theoretical work must be based on abstractions of the real world that are erroneous, overly simplistic, or both. Of course, the truth in most cases probably lies somewhere in the middle, and we are likely to make the most progress when the theoretical and empirical scientists work to inform and test each other in a mutually beneficial and open-minded manner.

In a similar vein, I wanted to find a way to apply my knowledge of and training in academic ecology and evolution to the world of restoration ecology in Hawai'i. I also hoped that working as a restoration ecologist might finally enable me to combine my love of rigorous empirical field science with my long-standing passion for on-the-ground conservation. While I did not regret that I had largely been obliged to put aside my applied conservation interests throughout my years in graduate school (at that time such interests were generally considered unworthy of the rigors of more formal academic science), I was looking forward to finally having a chance to try to accomplish both good science and real-world conservation.

So I eventually struck a compromise with myself: I would get my science fix via the formal research program proposed in our NSF grant application and by revising my dissertation papers and cranking out new ones, and I would get my conservation fix via the Ka'upulehu outplanting and by taking on a few more applied restoration and outreach projects. And maybe, with the right combination of luck, perseverance, and skill, some real conservation might trickle out of my more formal scientific research, and some real science might sneak its way into some of my applied restoration work.

Over the next several weeks, I divided the majority of my work time between revising our NSF grant application and thinking through and organizing the upcoming Ka'upulehu outplanting. We ultimately decided to pitch our proposal as a more focused investigation of the direct and indirect effects of removing key alien species from two diverse and degraded forested ecosystems in Hawai'i—the dry forest at Ka'upulehu and the mesic forest in Mahanaloa Gulch. We discussed how the direct effects of removing a dominant alien species such as fountain grass might include relatively obvious ecological responses, such as the reestablishment and spread of formerly suppressed native species and a decrease in the frequency and intensity of wildfires. We also pointed out the potential for various less obvious, unanticipated indirect effects, such as increased alien insect herbivory on rebounding native plant populations and the establishment of new noxious pests that also had been suppressed by exotic ungulates and dominant alien weeds.

In a nutshell, the essence of our argument, lovingly articulated in our thirty-three-page, single-spaced, ten-point-font grant proposal, was that although restoration programs frequently assume that native species will quickly and predictably recover following the removal of dominant alien species, few controlled scientific experiments had actually tested this hypothesis, particularly in heavily disturbed areas such as remnant native Hawaiian forests. We argued that since much of our knowledge of the ecology of alien invasive species comes from informal, anecdotal observations of the more obvious phenomena, our experimental, quantitative, and analytical approach would lead to a greater general understanding of the role of key alien species in the disruption of native communities. Finally, we maintained that our proposed research would yield practical information that could be of broad value to future restoration efforts within and beyond the Hawaiian Islands.

After the dizzying, on-the-ground crash course I had been receiving in the ecological and cultural complexities of life in Hawai'i, I enjoyed the more familiar intellectual challenges involved with helping to revise and polish that grant application (especially since Steve and Dave had already completed most of the hard work in their previous proposal). In contrast, working on the Ka'upulehu outplanting project was a humbling experience that continually forced me to confront just how little I knew about the myriad tasks associated with putting plants in the ground in particular and the broader practice of ecological restoration in general.

I had come to better appreciate both the ecological and political importance of this outplanting project. Several key players in the North Kona Dryland Forest Working Group (as well as in the Hawaiian conservation community as a whole) had expressed frustration with what they perceived as the NTBG's laissez-faire management at Ka'upulehu. Some had even suggested that it might be time to turn the lease of that preserve over to another group that would be more willing and able to actively restore it. Thus, regardless of how I and others might feel about its ultimate scientific value, I knew this project had become institutionally important for the NTBG as a whole.

"Okay," I told myself, "if I have to spearhead this project, I might as well do it right by utilizing and applying all my scientific knowledge and training . . . but how exactly do I do that?" After all those years of reading and even writing about how valuable rigorous ecological science could be and already was to practitioners, I suddenly found myself on the other side of this fence. With some chagrin, I recalled how often we academics complained about the fact that so many practitioners apparently didn't read our scientific literature, didn't understand the fundamental principles of ecology and evolution, and so on. However, now that I was working with and becoming friends with some of these people, I was starting to better appreciate the complexity of their work, the hectic nature of their lives, and the extent to which much of this literature was inaccessible, unintelligible, or seemingly irrelevant to the larger practitioner community. But I read and even published in that literature and I had my PhD in ecology; shouldn't I know or be able to figure out how to scientifically implement this simple little outplanting project?

Yet the more I delved into it, the more complex and ambiguous it all became. Rather than concrete answers, I found only swirling cascades of new issues and questions I hadn't even considered. And while the whole idea of and motivation for this project had at first seemed straightforward, I now felt baffled by even the most basic and simple questions. For example, where exactly should we transplant all those plants within the Ka'u-pulehu Preserve—in the shade of the existing tree canopy or out in the open? Within the deepest soil pockets or the mostly soilless cracks and crevices in the 'a'ā lava outcrops? How should we plant them—pack them together in dense clusters or give each plant lots of space? Which species should be interplanted and which should be segregated? And what, if any, standard horticultural practices should we employ—spray each plant with insecticide and dip its roots in rooting hormone prior to transplantation and then fertilize and water them after outplanting, or just throw them into the ground and let Darwinian selection decide which, if any, are fit enough to survive?

As I wrestled with these questions, I soon found myself pondering perhaps the most basic, yet deceptively simple-sounding, question of ecological restoration projects in general: What historical time period and particular ecosystem within that time period can and should serve as our "pristine" reference model?

Resolving this more conceptual issue can be fiendishly difficult and contentious because the "answer" may hinge more on personal values and philosophies than on objective science. Ironically, much of the ambiguity and subjectivity surrounding this question arise from the modern non-equilibrium model of nature. Since the mid-twentieth century, ecologists have increasingly rejected the older, more static and orderly view (ecosystems progress through a series of predictable successional stages until reaching their stable, "climax" state) in favor of a more dynamic and chaotic paradigm (ecosystems and the biosphere as a whole are complex entities that can change in unpredictable ways, and even without human-caused disturbances they may never reach a stable, equilibrium-like state). Thus, if nature is an ever-changing and stochastic beast, then the selection of specific historical ecological restoration targets is more a matter of taste than of science.

In the case of native Hawaiian dry forests, our current scientific understanding is that their precontact distribution and diversity appear to have been the product of physical (e.g., lava flows and weather patterns) and biological (e.g., species colonizations and soil development) interactions that resulted in a complex spatiotemporal mosaic of forest fragments with different substrate ages and species compositions. In fact, we now believe that even before humans appeared and began burning and clearing these forests, phenomena such as catastrophic volcanic eruptions, changing climatic patterns, and the arrival of new species may often have produced both subtle and radical ecological changes within this ecosystem. Depending on which prehuman time period one chooses, the "pristine" ecological reference model for a given degraded chunk of land in Hawai'i today could range from a diverse dry forest or shrubland all the way to a low-diversity rain forest.

Once humans enter the equation, the task of selecting a historical ecological reference point becomes even more murky and subjective. As is the case for most of the world's ecosystems, the distribution and structure of Hawaiian dry forests in the postcontact era have increasingly reflected the direct and indirect effects of human disturbances such as deforestation, fire, and invasions by exotic species. Today, the interaction between these human-mediated effects and precontact biological and physical phenomena has produced a system (wreckage?) of disjunct, degraded dry forest fragments with largely unknown ecological dynamics and ecosystem processes.

About the only thing that was clear to me was that regardless of how these dry forests used to work, today they were tragically broken. Among the obvious kinds of damage were native species extinctions, alien species invasions, and habitat destruction and degradation; perhaps less obvious were massive deforestation, possibly leading to a hotter and drier climate, and loss of essential soil nutrients that were formerly transported from the ocean to lowland terrestrial habitats by the vast precontact flocks of birds. The net result, not surprisingly, was that the only way some semblance of this ecosystem could survive was through deliberate, aggressive, and extensive human interventions. Yet given the myriad irreversible changes to these ecosystems, the fluidity of their past, our ignorance of even their basic "natural" ecological processes such as succession, and the unpredictability of the future, what exactly should or could we do with those transplants at Ka'upulehu?

"Well," I finally told myself one day when I was just about ready to abandon the whole project, "maybe I should just lighten up a bit!" If we don't and probably never will know much about the history of Hawai'i's dry forests except that they were dynamic and changing, but we do know that we could never restore that Ka'upulehu parcel to anything approaching its former grandeur even if we knew exactly what that once looked like and how it functioned, why don't we just try a few plausible things and see what, if anything, might work now? Moreover, given that there apparently was not a single even semifunctional dry forest remnant left anywhere in the entire Hawaiian archipelago that we might use as a contemporary reference point (the so-called vanishing baseline that unfortunately has become increasingly common in restoration ecology), who could say that whatever we did was wrong?

Feeling more hopeful, I shifted my focus to yet another basic, seemingly simple ecological restoration question: Who is this project for? This too had seemed straightforward at first. The Ka'upulehu Preserve is a fragment of a globally endangered ecosystem, with a handful of federally endangered plant species. If we think that Hawai'i's native biodiversity is worth saving, and that human-caused extinction is something we should try to prevent, then transplanting all those native plants back into the "wild" is a no-brainer.

Yet even though I was still getting to know the various individuals and organizations that made up the North Kona Dryland Forest Working Group, it was already clear to me that within the group there were some very different personal and institutional reasons for attempting to restore Ka'upulehu and "save" Hawai'i's native dry forests in general. For example, even though they might not necessarily come right out and say it, I knew that some in the group were less interested in preserving Earth's biodiversity per se than in cultural history, tourism, and public relations (e.g., showing tourists all the wonderful things the real estate and resort industries were doing for the land and people of Hawai'i) or in potential local environmental benefits (e.g., finding ways to control fountain grass, reduce the size and frequency of wildfires, and increase local rainfall by reforesting all that grassland). And although I happened to care deeply about maintaining species diversity and preventing extinctions, I had both personal and institutional reasons to focus primarily on the scientific aspects of this restoration project.

As I thought about it more, I realized that at least as I had originally envisioned it, the immediate answer to my question was that this project was for the US Fish and Wildlife Service, as it was the only entity in our working group explicitly mandated to preserve the biodiversity of the United States. Representatives from this agency had also played key roles in forming this group and providing the necessary funding to get it started and keep it going. Nevertheless, envisioning the project as being solely for this agency seemed disingenuous. As much as I liked and respected the

USFWS representatives I had met, like me they were all relatively afflu-
ent, educated white people from the US mainland. They didn't live on the
Big Island—except for those who worked within the agency's system of national wildlife refuges, all of Hawai'i's USFWS employees lived on O'ahu
and worked out of the Honolulu office—and they were not members of
the local community who, one hoped, would continue to care for and benefit from this restoration project long after we *haoles* were gone.

On the other hand, given the substantial involvement of so many nonlocal individuals, agencies, and organizations, it would be equally misleading to think of this as a purely local, grassroots restoration project.
Right or wrong, I seemed to be the one organizing the whole thing and agonizing over all these philosophical, ecological, and practical issues.
Moreover, I knew that there was no discrete and homogenous "local community" in North Kona to consult, even if I could fly back there and hold
some type of public meeting, because "the locals" who lived in that region
represented, like people in general and throughout the Hawaiian Islands
in particular, a complex mixture of different ethnicities, interest groups,
and socioeconomic classes.

Indeed, over time I discovered that within North Kona, and throughout the state as a whole, the Hannah Springers and Michael Tomiches
were the rare exception, not the rule. The majority of people who lived on
or visited the leeward side of the Big Island knew little if anything about
the plight of Hawai'i's native species, let alone the past and present state of
their local dry forests, and most would probably have been too busy or dis-
interested to attend a meeting on these subjects. Sure, some might know
and care about the particular native species traditionally important to
something they themselves enjoyed doing, such as hula dancing or canoe
racing. And there might be a handful of locals (actual natives as well as
haole transplants) who were into all things Hawaiian, although, unlike the
USFWS, they might well value the nonnative, Polynesian-introduced species more than some of the native, presently endangered but culturally
unimportant species we would be transplanting. But on the whole, just as
with everywhere else I have lived, I was willing to bet that if push came to
shove, the majority of North Kona's residents would choose to replace *any*
undeveloped land, regardless of its biological or cultural value, with yet
more houses, resorts, businesses, roads, parking lots, ranches, and golf
courses.

"Well . . . ," I finally told myself, feeling lost once again, "maybe, like
everything else, the question of 'Who is this restoration project for?' also
has no clear and satisfying answer. Maybe I should go back to the 'lighten

up' philosophy, roll up my sleeves, and just start doing *something* before it's too late!"

Once I finally got down to planning the nitty-gritty details, I found myself increasingly consulting practitioners such as field technicians, staff gardeners, and irrigation specialists. Although I discovered that the "right" way to proceed in this arena appeared once again to be a complex function of subjective philosophy (baby the plants initially to compensate for the artificially harsh conditions at Ka'upulehu today or employ a tough love approach?) and personal histories (agricultural, horticultural, tourism industry, or conservation background and orientation; formative experiences in wet or dry environments?), at least these people could speak firsthand about such topics as their successes and failures with various soil amendments and watering regimes and the relative merits of the Pulaski ax and the 'ō'ō, a Hawaiian digging bar traditionally constructed of wood from the hardest dry forest trees.

Interestingly, I found that the more practical, on-the-ground experience people had, the more they tended to view their work as an idiosyncratic art rather than generalizable science. The old hands would warn me that a planting and watering regime that appeared to work well for, say, a large potted specimen of a fast-growing shrub transplanted into a moist, shaded, deep-soiled forest might prove to be disastrous for a small, slow-growing canopy tree outplanted into the austere environment of a barren lava outcrop at Ka'upulehu. Several of these people suggested that the best approach was to conduct a series of informal trial-and-error experiments, closely watch what happened, and then go with what seemed to work best for each combination of species and outplanting site. While this was probably sage advice, given our time, money, and logistic constraints plus my desire to incorporate some rigorous science into this restoration project, I decided to standardize all our horticultural procedures and err on the tender-loving-care side of the philosophical continuum.

Once my plan was sufficiently detailed and concrete, I realized there was no way even a small army of volunteers and I would be able to get all those potted plants into the ground at Ka'upulehu in one day unless a substantial amount of work was completed beforehand. Dave, Tim, an NTBG irrigation specialist, and I therefore flew back to the Big Island before the outplanting day to complete as many of these preplanting tasks as possible.

Much as with my own more academic, pure science experiments in the deserts of New Mexico, being in the field and trying to visualize and plan our operation forced me to almost immediately revise the plans I had carefully formulated back in the office. Similarly, many of the theoretical and philosophical conundrums I had agonized over on Kaua'i ended up being somewhat ingloriously resolved by mundane practical considerations.

The first and probably most important logistic issue I had failed to anticipate was that every outplant was going to need its own irrigation line. Several experienced people had convinced me that without regular and substantial supplemental water, at least until the outplants were established and recovered from transplant shock, few if any of them would be likely to survive for long. Yet although the manager of the adjacent Hualalai Ranch had kindly run an irrigation line from his cattle troughs down to the upper corner of the Ka'upulehu Preserve, this gravity-fed tube could carry only a modest amount of water at fairly low pressure. Consequently, the only viable option was to run slow, water-conserving, pressure-compensating drip irrigation emitters to each and every plant, as any kind of area sprinkling system would have required far more water and pressure than we had. In fact, after some initial tinkering, we found that even with this system, there would not be enough pressure to irrigate all the plants at once, and thus we would have to install a timer and sequentially water a series of smaller blocks of plants.

Our original plan had been to spread the roughly 400 plants in the NTBG's nursery slated for this project evenly across the sixteen ten-by-ten-meter plots we had so laboriously hand-cleared of fountain grass on my first trip to Ka'upulehu. But as I walked around the exclosure and tried to imagine running all those irrigation lines and timers and emitters to all the plants in each of the sixteen plots scattered across the preserve, then conducting regular post-outplanting data collection censuses, I smelled disaster. So, after much walking around, experimental diggings, and back-and-forth discussions, we finally agreed on a much simpler plan: (1) Establish four new outplanting sites spaced at roughly uniform distances across the preserve. (2) Put two of these in the shade of existing tree canopies and two out in the open, sunny intercanopy areas. (3) Plant 50 plants at each site, for a grand total of only 200 plants; this would also allow us to select the best specimens of each species and not waste our time and resources with the sick and dying ones.

The sun-and-shade treatment reflected my judgment that available light was the single most important variable we could meaningfully investigate in this outplanting project. My rationale was that although we probably would never know the specific niches in which these dry forest species used to germinate and establish themselves, perhaps we could discover the optimal light levels for them in today's altered and degraded environment. Ideally, I hoped, a few years down the road we might have the data to conclude that "species A and B apparently need a shaded environment to establish, species X and Y need full sun, and it doesn't matter for C, D, and E."

Despite my effort to make this experimental component of the outplanting project as clean as possible, I knew we would very likely face a barrage of criticisms if and when we tried to publish our results in a decent scientific journal. This was my first of what would be many experiences of the challenges associated with trying to simultaneously design and implement scientific experiments and applied restoration projects. In this and many other cases, much of the tension stemmed from the need to balance the often conflicting demands of the science of restoration ecology for such things as rigor (e.g., carefully controlled and replicated treatments) with the practice of ecological restoration's demands for such things as efficiency (getting the highest number of plants in the ground as quickly and cheaply as possible).

For example, some scientists might reasonably argue that in this case, our independent variable (sunlight) was confounded by a suite of other factors that we did not control or measure. That is, any differences that we detected in the survival and growth of the plants in the full-sun plots relative to those in the shade could have been caused by other, nonlight variables such as different underlying soils, relative humidities, and densities of existing vegetation.

Some could also point out that we had not even quantified and standardized our experimental variable. I could almost hear these critics already: How many photons of light did your "full-sun" and "shade" plants receive? Were the light levels in the two replicates of each treatment level really equivalent? Did all the plants in each of the four outplanting sites receive the same amount of light, or was there substantial intrasite variation?

Yet another major category of criticism we might justly receive could focus on the amount and type of our replication. The statisticians would not like the fact that there were only two replicates for each level of our experimental treatment. Similarly, I knew there were some sticklers who would argue that because our study took place in the Ka'upulehu Preserve, the entire experiment was pseudoreplicated, and thus technically our results could not and should not be extrapolated beyond this particular exclosure.

In a nutshell, the sin of pseudoreplication involves treating data as independent when in reality they are interdependent. For example, if I measured the sizes of 10 leaves growing on the same oak tree, I could not technically claim that I had 10 independent measures of leaf size; what I should claim is that I had 10 replicate samples of the leaf size on that particular oak tree. If I wanted to rigorously test some hypothesis about the size of oak leaves for the forest in which that oak grew, I would need to measure leaves from many different trees in that forest. In other words, there is a critical statistical and ecological difference between measuring 1,000 leaves from a single tree and measuring 10 leaves from 100 different trees. Doing the former and statistically handling those data as if they had been obtained in the latter sampling design would be a classic case of pseudoreplication.

While this hypothetical example represents an unambiguous and extreme instance of pseudoreplication, ecologists and statisticians often reach markedly different conclusions about what does and does not constitute pseudoreplication in practice. Similarly, editors and the expert scientists they employ to review their manuscripts often have very different tolerances of and sympathies for the kinds of real-world problems that can force applied scientists to utilize less than ideal methodologies, such as potential pseudoreplication (there was only one native dry forest remnant available for this project—sorry!); too few statistical replicates (given our limited water and pressure, it was miraculous that we managed to irrigate four separate plots); or messy experimental treatments (we didn't have the time, money, or equipment to quantify and standardize the light levels; this was really a restoration project, so give us a break!).

Once we finally selected our four outplanting areas—again mostly on the basis of practical considerations such as the logistics of irrigation and where we could dig sufficiently large holes—it didn't take long for us to see that in fact there were some other potentially important intra- and intersite differences. Perhaps the most striking of these differences was that the soil was in general much better and deeper under the trees. (Was this because these trees had differentially colonized and established the best soil patches in the first place, or had the trees themselves improved or conserved the soil that was already there?) Another obvious nonlight difference was temperature: a few hearty swings of my pickax were more than sufficient for me to realize (duh!) that it was hotter out in the open, especially when I was standing on or near exposed black lava.

While scientists tend to focus their criticisms on one another's experimental designs and quantitative analyses, I have often found that the inherent heterogeneity of nature may actually create more important and fundamental problems for the discipline of field biology in general and restoration ecology in particular. Just as I had eventually discovered in the New Mexican deserts, once I got down on the ground at Kaʻupulehu and really started to look, I noticed how radically this seemingly homogenous forest fragment could change across even small spatial scales. For instance, the composition, size, and vigor of the different plant species might change completely, or the fountain grass coverage could transition from a dense, impenetrable blanket to a thin covering of a few wispy individual stems. Similarly, once I started digging the outplanting holes, I often encountered a fine-scale mosaic of deep pockets of rich black soil, soilless patches of gravelly rocks, and sheets of bare ʻaʻā lava. Although these kinds of intrasite, intratreatment differences can be critically important to applied practitioners, they can be difficult if not impossible to detect and standardize with even the most rigorous scientific methodologies, equipment, and statistical analyses.

In graduate school, I had increasingly noticed and been troubled by what I felt were some major gaps between the underlying assumptions of ecological methodologies in theory and what I was seeing in nature, but because I did virtually all of my fieldwork myself, I had not had to deal with the problems associated with fine-scale *human* variability. But at Kaʻupulehu, once we started digging it was immediately apparent that each of us dug our holes differently. Even though we strived to standardize the dimensions of all the holes, differences in the way each of us responded to the natural variability we encountered sometimes resulted in substantial differences among our holes. For instance, what do you do when you dislodge a big rock and it leaves behind a crater bigger than the target hole depth? How do you stabilize the sides of a hole dug largely out of a gravel pit?

While I knew that in theory we could attempt to control for this variability by evenly distributing our holes within and among each of the four outplanting areas so that no one site, or patch within a site, had a disproportionate number of holes dug by a single person, in practice this strategy would have been inefficient and impractical. Some of us were better and faster diggers (Tim was a maniac); sometimes it made more sense to split up and work in different outplanting areas; some of us were occasionally needed for other tasks; and so forth. Similarly, while we could have measured each completed hole and reconfigured the ones that deviated too far from the target dimensions, this would have required an enormous amount of our limited time and energy and probably would have sent us back down the infinite loop of differential human responses to fine-scale natural variability: Where exactly do you place your ruler when the bottom of the hole is deeply concave? How do you handle holes that aren't round?

Finally, given that our plants would be transplanted by volunteers with widely varied levels of skill and dedication, attempting to achieve that level of precision would probably have been misguided anyway. Thus, in the same way that we most likely would never resolve many of the theoretical and academic issues surrounding this restoration project, I began to realize that no matter what we did, the combination of uncontrollable natural and human variability was going to result in the kind of messy ecological field experiment that I had been trained to judge as inferior to the far more rigorous research programs typically implemented by purely academic scientists.

"But this is really a restoration project with a little scientific experiment serendipitously tacked on," I tried to console my increasingly polarized self. "We will be lucky if we even manage to get all our holes dug and the irrigation system set up before we have to fly back to Kaua'i. Every additional step we take to tighten up our methodology is costing us precious time; if we don't get everything done now, both the scientific and restoration components of this project will suffer. So once again, lighten up! Maybe this admittedly flawed experiment will one day provide someone with useful information, even if these data are ultimately rejected by the gatekeepers of professional science." Although I didn't know it then, this internal dialog was merely a mild prelude to what, over the course of my career as a restoration scientist and practitioner, would become a cyclic internal battle to fend off a creeping wave of schizophrenia.

After three long, hard days, which yielded another round of sunburns, blisters, and vows to get in better shape, we somehow managed to get it all done. Now all I had to do was plan and implement the last layer of project logistics, which of course involved yet another suite of tasks for which I was completely inexperienced and untrained: How best to package and ship all those plants on Kaua'i to the Big Island? How many volunteers do we need to get all 200 plants in the ground in one day, and how do we get them? What happens if someone gets hurt? "When this whole thing is finally over," I promised myself, "this time I'm really going to spend some quality time on the beach."

Compared with all the mental and physical stress of preparing for it, our outplanting day was a dream. When Dave and I flew back to the Big Island the following week, we were met at the Kona airport by a lovely elderly couple who had thoughtfully rounded up a van for us to transport our plants and equipment up to the Ka'upulehu Preserve. I was immediately struck by their enthusiasm, knowledge, and kindness—they even brought a cooler full of drinks and homemade food!

I was similarly amazed and inspired by the number and diversity of people who showed up bright and early the next morning. This group included members of the North Kona Dryland Forest Working Group, local firemen, staff from some of the nearby resorts, and an eclectic assortment of other Big Island residents. Several of these volunteers had also brought their kids, most of whom seemed to grasp the significance of this project and were genuinely excited to be a part of it.

As the morning progressed, I couldn't help noticing how different we all were. In almost any other situation, most of us would have little if anything to say to one another, and if for some reason we did strike up a substantive conversation, we probably would have discovered that we had radically different opinions about such things as politics and religion. Yet here we were, donating our time on a beautiful Saturday morning and working harmoniously together.

Was there something about the experience of getting one's hands in the soil and planting that helped us unite and set aside our differences? Could we have been under the spell of some powerful spirit that still haunted the remnants of that once mighty Ka'upulehu forest? While I tend to be skeptical of such things, I must admit that as soon as Hannah began chanting in Hawaiian and calling out to her ancestors, even though I didn't understand a word of it, a wave of spine-tingling goose bumps (or "chicken skin," as the Hawaiians call it) swept over me, and I choked up later when I saw her and Michael's two young, beautiful children reverently transplanting *lama* saplings.

By eleven that morning, all 200 plants were safely in their holes and, we hoped, soaking up their individual trickles of water. After everyone left, I went around and did a final quality control census. Much to my pleasant surprise, I found that virtually all of the plants had been carefully tucked into their holes, and all of my numbered aluminum tags had been wired securely to their designated adjacent stakes. With a few mouse clicks back in the office, each of those numbers could reveal the history of its plant— when the seed that produced it had been collected and by whom, which tree within this exclosure was its mother, its height and width at the time of outplanting, and so on. It was sobering to realize that it would probably take several hundred years for these saplings to grow into what could legitimately be considered mature canopy trees. If at least a few of them did survive that long, which if any pieces of data would the people of the future find most interesting and important, and what kinds of information might they wish we had provided?

Even if this whole project was misguided or none of the plants survived, it sure felt good to plant them, and it was immeasurably inspiring to walk around and see them in the ground. No matter what happened, at least we tried. As I shouldered my pack and headed down toward the highway, it occurred to me that this little scrubby, basket-case dry forest remnant was really starting to grow on me.

Chapter 3

Now What? Responding to Nature's Response

I parked my rental car on the gravelly lava shoulder between the Mamala-hoa Highway and the Ka'upulehu Dry Forest Preserve and tried to rub the sleep out of my eyes. As excited and curious as I was to see how all our plants were doing (it had been six months since that outplanting day and three months since my last census of them), I also yearned to lie down in the backseat and take a nap. But, as always, there was far too much to do and far too little time in which to do it before I had to fly back to Kaua'i on yet another red-eye flight, so I fought off the urge to sleep, slathered myself with *haole* war paint (a.k.a. sunscreen), and rolled on out of the car.

While the ultimate scientific and conservation value of those outplants remained unclear, they had proven to be an extremely effective tool for public relations and community building. Whether it was the sheer audacity of attempting this project in such a degraded ecosystem (just about all the learned people I knew in Hawai'i thought we were nuts) or simply the experience of getting down and dirty together, the process of putting all those plants back into their ancestral Ka'upulehu soil had somehow inspired and unified our North Kona Dryland Forest Working Group in a way that all our previous meetings had not.

Moreover, largely because of that outplanting project, more and more people were coming out to see for themselves what we were up to. And with the once mighty fountain grass largely vanquished, even relatively timid members of the public could now walk through much of the exclosure, see and touch the gnarled old endangered trees, and experience being in a "real" native Hawaiian dry forest. Even better, when our visitors encountered the brazen patches of transplants within our spaghetti-like maze of irrigation lines, many started asking good questions, and some wanted to know how they could help.

When I reached the first outplanting site, I was relieved to see that the plants appeared to be in good health and the irrigation system seemed to be functioning properly. We'd had an annoying series of problems with it at first—the timers not switching on and off when they were supposed to, too much water going to some sites and not enough to others, ants clogging up the lines. At one point the system somehow stayed on and drained away thousands of gallons of water—until the manager of Hualalai Ranch, Franklin Boteilho, stormed down and turned it off after discovering that his tanks, which fed our system, were empty and his cattle troughs were dry.

On a previous trip, I had driven up to Hualalai's headquarters to meet Franklin and ask his permission to access the ranch's lands surrounding the preserve for research purposes. Like most of the other ranchers I'd met, he was a no-nonsense, hardworking man with little time for or interest in small talk. It had been a major coup, and a testament to Hannah Springer and Michael Tomich's diplomatic skills and general standing in the local community, to get Franklin to even come to one of the working group meetings. Nevertheless, he immediately let me know in no uncertain terms where he stood and how he felt. He was also still ticked off by the amount of water we'd wasted and viewed that incident as just another ex- ample of the blundering incompetence of environmentalists, scientists, government agencies, and clueless *haoles*. "Every time I give you guys an inch," he railed at me with outstretched arms, "you screw it up, then come back and want a mile!"

I think what typically most angers and hardens people like Franklin, who have acquired their knowledge and skill through direct experience and sweat, is the often justified perception of being ignored and even ridiculed by those with advanced degrees after their names. Yet, as I expe- rienced repeatedly throughout my time in Hawai'i and elsewhere, once I let the Franklins of the world get their pent-up diatribes off their chest, they tended to calm down and turn out to be surprisingly reasonable and insightful people. In this case, I ended up learning a lot of useful and in- teresting information from him about the cultural history and ecological past and present status of the North Kona landscape. While we may have had fundamentally different visions for Ka'upulehu's future, at least we were able to understand and appreciate each other's point of view.

I followed the irrigation line down to the first hole in the first outplanting site and checked the plant's tag: "A1: *Colubrina oppositifolia*." A member of the Buckthorn family, this was another federally endangered dry and mesic forest tree that had become increasingly difficult to find both on the leeward side of the Big Island and in its only other known extant location, in the Wai'anae Mountains on O'ahu. It produces exceptionally dense, hard wood that Hawaiians once used to make *kapa* (cloth) beaters, vicious spears, and poles for construction.

Interestingly, the Hawaiian name for this species, *kauila*, also refers to *Alphitonia ponderosa*, another rare endemic tree in the Buckthorn family that occurs in dry and mesic forests. This kind of botanical ambiguity within their language is unusual, presumably because the Hawaiians were keen observers of the natural world and paid particularly close attention to the species that had utilitarian value. Some have speculated that the word *kauila* referred to the timber of these two trees rather than the species themselves; *Alphitonia* also produces an extremely dense, heavy wood, and ancient artifacts constructed of these two species are apparently indistinguishable without destructive sampling. In any event, I eventually discovered that some of our more learned young male visitors loved getting photographed while striking a macho warrior pose in front of what we called our "grandma" *kauila* tree.

I fished out my Outplanting Site A data sheet and found the row for Hole #1. Looking back across the columns of data, I saw that this particular seedling, like all of the forty other *kauila* trees we had transplanted here, originated from that grandma tree. There were ten individual *kauila* trees within the Ka'upulehu exclosure, but grandma was by far the largest as well as the most prolific (and often only) seed producer. As much as we would have liked to increase its local genetic diversity by planting seedlings derived from many different and unrelated maternal genotypes, we simply had no choice because grandma's seeds were all we had.

In theory, we could have queried other botanists across the state to see if anyone had seeds of this species that we could use in our outplanting project. In practice, however, I knew that even if we had been able to get our hands on *kauila* seedlings produced from non-Ka'upulehu seeds, attempting to transplant them into the preserve could have led to lengthy and divisive bureaucratic and political battles.

The worlds of conservation and restoration are composed of individuals and agencies that form a continuum running from conservative "purists" to pragmatic tinkerers to laissez-faire, try-anything "artists." As I discovered while trying to conceptualize and implement this outplanting project, Hawai'i appears to have more than its share of purists who fear making things worse than they already are. Given the islands' tragic ecological history, and the occasionally disastrous consequences of actions undertaken even by careful, well-intentioned people, this philosophical position is certainly understandable. The members of this camp would thus have vehemently argued that bringing in species from outside our unique Ka'upulehu bioregion could destroy its ecological and evolutionary integrity. They would also have pointed out that those foreign *kauila* trees would most likely eventually mate with the existing Ka'upulehu specimens—we certainly hoped they would—and potentially contaminate their "locally adapted gene complexes" with novel genes that could prove to be poorly suited to this particular ecosystem (so-called outbreeding depression).

The pragmatists in the middle of this continuum might agree that while we should always carefully think through the consequences of our actions and design and implement our programs with as much care and rigor as possible, we must also consider the risks of *not* doing something. For example, in our situation, it is plausible that the genetic diversity of the Ka'upulehu *kauila* population is already dangerously low, whether from past inbreeding (all of the trees within the exclosure may be close relatives), from restricted gene flow due to such factors as the potential loss of native pollinators and asynchronous flowering among the few reproductively mature individuals in this area, or both. Thus, not increasing its genetic diversity by bringing in unrelated individuals with different genotypes could decrease this population's ability to adapt to its changing local environment and contribute to its eventual local or even global extinction.

When it comes to this debate about local adaptation versus genetic diversity, most purists and pragmatists would agree that we should try to hit both targets by striving to find seeds produced by trees that grew within this particular ecological environment yet were not closely related to one another. But what should we do when, as is so often the case with endangered species such as *kauila*, this isn't a viable option? The Big Island's remaining *kauila* population is in fact small and scattered, and even if we had the time, resources, and permission to roam around the island searching for ripe fruits, how would we decide whether the seeds (assuming we ever found any) from a particular maternal tree were produced in a sufficiently "similar" ecological environment? How would we know how genetically similar the seeds matured by a *kauila* tree growing, say, fifteen miles away were to those produced by our grandma?

A purist might argue that before proceeding we should make every effort to answer such questions using the best available science. In this case, we could collect relevant ecological information for any area in which we found additional *kauila* seeds (its species composition, underlying substrate type and age, temperature and precipitation patterns, etc.) and then compare these data with the biotic and abiotic habitat of the Kaʻupulehu Preserve. Similarly, we could perform genetic analyses of all the new seed-producing trees we found and compare these data with the genetic structure of the extant *kauila* population within the Kaʻupulehu exclosure.

However, if we had the ability and expertise to collect and analyze all this additional information (a situation that rarely if ever would occur in the real world), even these data would not necessarily resolve the relevant disagreements. No matter how much quantitative data have been collected, people can and do argue about what constitutes a sufficiently similar ecological environment: Which is more important, a habitat's substrate age, soil depth, or species composition? Is a difference of five centimeters of annual precipitation or 500 feet of elevation too much? And the analysis and interpretation of ecological genetic data tend to be even more murky and contentious. Moreover, if such studies suggested that relative to grandma *kauila*, the genetic relationship of seeds collected on Oʻahu was not significantly different from the relationship of seeds collected from trees growing five miles away from our grandma, most purists would point out the limitations of such testing and still argue against bringing plants derived from the Oʻahu seeds to the Big Island.

Even within the pragmatists' camp, there is often a wide range of philosophies regarding what we should and should not do. Some would say that when it comes to an endangered species such as *kauila*, if it is difficult or impossible to get a sufficient number of seedlings, with a sufficient amount of genetic diversity (the de facto definition of "sufficient" in this context being "study the particulars of your situation carefully and use common sense"), after a "reasonable" amount of local searching, then by all means go ahead and use seeds collected from other ecosystems on the Big Island, or even from the other islands if necessary. We know little about how much gene flow there was within and among the islands before humans first arrived, or how much the Hawaiians consciously or inadvertently moved plants and animals beyond their prehuman geographic and ecological boundaries. So maybe, in the end, bringing in *kauila* seeds from Oʻahu would be both a necessary and "natural" thing to do.

From this more liberal perspective, the Hippocratic oath of medical doctors to "first do no harm" is a misguided restoration philosophy because in these kinds of dire situations, playing it safe for fear of making things worse is ultimately a recipe for failure. As one of my colleagues put it, "If we are going to use medical analogies, a better one would be to think of a triage situation in an emergency room. No matter what we do, some people [or species and their ecosystems] are going to die, but if we are afraid to take chances and do nothing, we'll almost certainly lose everyone! Better to give it our best shot and try to help as many as we can, even if in the end there were a few individuals who would have fared better if we'd left them alone."

Yet I know many other restoration ecologists who consider themselves pragmatists but see this philosophy as leading down a dangerously slippery slope. Their take on the triage analogy is that first, only highly trained and experienced doctors (scientists) should be allowed to make decisions in these kinds of situations—we wouldn't let people run in off the street and attempt a bunch of heroic maneuvers to try to save everybody, no matter how noble their intentions might be. Second, even the most accomplished medical professionals must abide by an established set of rules and procedures that have been carefully designed for these kinds of situations—the emergency room is not the time or place for experimentation! And third, there are explicit guidelines for prioritizing treatments and resources among patients in triage situations—radical and costly treatments would not be administered to one desperate patient if that meant neglecting five others who could have been saved by relatively simple and cheap techniques. Members of this more conservative camp thus might be willing to concede that bringing in plants or animals from other parts of an island is warranted in some cases but might draw the line at going off-island to get them. Some of these people would also question the wisdom of working in an ecosystem as desperate as Hawaiian dry forests in the first place!

Finally, at the other end of the continuum are the "heretics" who believe that the whole paradigm of historical (or "hysterical," in their eyes) ecological restoration is both misguided and futile. To them, issues such as prehuman ecology and evolution, gene flow and local adaptation, and even the human-nature dichotomy are irrelevant; some think the very idea of "wilderness" is both a meaningless, imaginary human construct and an oxymoronic resource management objective.

There is also a substantial amount of philosophical diversity within the heretic camp. Some essentially believe that Earth is a canvas and we are the artists, and thus we can and should feel as free as a painter to mold "nature" into whatever we want it to be. Others think that we should orient our management practices toward achieving various ecological objectives such as maximizing biodiversity, ecosystem stability, and utilitarian services such as water conservation and food production. When it comes to

Hawai'i, there are those who argue that rather than wasting time and money trying to save the archipelago's doomed native flora and fauna, we could be turning these islands into a sort of Noah's ark of the world's vanishing tropical biodiversity. (There are in fact some species that are now rare or degraded in their native countries but flourishing in Hawai'i.) There is considerable sympathy, if not outright support, for these kinds of proposals among various sectors of Hawai'i's business community and within the public at large. And when we "environmental Nazis" demand ever more resources to kill all those cats and rats and frogs and ungulates and gorgeous flowering plants, even some of the more ecologically informed people and organizations start to wonder whether we really are fighting the good fight.

I finished my census of the first outplanting site and paused to eat what had become my standard field breakfast, Hawaiian apple bananas dipped in poi (mashed taro root). As intriguing as all these intellectual labyrinths and ethical dilemmas could be, I increasingly found myself just focusing on the relatively mundane practical issues that seemed to perpetually pop up around me like noxious weeds. Yet trying to think through these practical issues often led me right back into the philosophical side of ecological restoration. For example, we regularly observed that the leaves on many of our outplanted *kauila* and *koki'o* (*Kokia drynarioides*) seedlings were discolored, wilted, or eaten by some unknown herbivore. Since we had never seen any wild seedlings of these species, we didn't know whether these things were potentially problematic or "normal." Similarly, I noticed that some of our sandalwood seedlings were chlorotic, and a few were already dead. Could this be caused by the absence of a suitable host plant for their obligate root parasitism? What species did they once parasitize? In prehuman, intact dry forests, how long had it taken the roots of wild sandalwood seedlings to find their hosts—maybe they are supposed to look this way at first? Could we, should we attempt to solve this problem by applying fertilizer or by planting nonnative but effective "host root" species nearby, as apparently some of the nurseries did?

I was also busy trying to detect and tease apart any general patterns that might help inform and guide our larger dry forest restoration program at Ka'upulehu. For instance, although there was considerable variability among the different species, it appeared that at least the initial growth and survival of the outplants as a whole were greater in the shaded sites. Once again, this observation raised more questions than it answered: If real, what was the causal mechanism behind this result—the reduced light itself? Better soil? Should we attempt to test one or more of these hypotheses by, say, erecting shade cloth structures over the full-sun outplants, or would it be better to start over and design a "real" scientific experiment?

I finished my breakfast, took one last admiring look at all the outplants, and then headed out to attend that morning's working group meeting. With a little distance, a selective field of view, and some imagination, it almost looked like a healthy, regenerating dry forest—some of the faster-growing shrubs had already started to flower and fruit. I wondered when a human being had last seen such a diversity and abundance of native Hawaiian dry forest seedlings and saplings "in the wild" on this island, or, come to think of it, anywhere within the entire archipelago.

I kicked off my sneakers and added them to the pile of assorted boots, dress shoes, and flip-flops that lay outside the side entrance of Hannah and Michael's house at Kukuiohiwai. I loved their home—it was full of fascinating Hawaiian artifacts that had been passed down through their families for generations, it had breathtaking ocean views, and it somehow felt both spacious and cozy at the same time. It was also a welcome relief from the cold and sterile environments of the institutional buildings in which I seemed to spend an ever-increasing portion of my life.

I stepped in and was startled to see how crowded it was—I didn't even recognize several of the dozen or so people I counted around the room. We had slowly but surely become an alphabet soup of institutions and individuals: our working group list now included a broad array of local individuals, civic groups, nonprofit agencies, for-profit businesses, environmental groups, botanical gardens, universities, and a slew of government agencies ranging from the US Army to the Department of Hawaiian Home Lands.

After an opening round of personal introductions, our meeting facilitator, Andrea Beck, from the Hawai'i Forest Industry Association, distributed the minutes of and reviewed the pending assignments from our previous meeting, two months earlier. As usual, we seemed to be trying to juggle a few too many balls at once. Our current business included the following:

1. Reports. Topics requiring written documentation—for various external agencies, our own internal purposes, or both—ranged from the general ecology of tropical dry forests to fountain grass biomass and flammability to summaries of our past accomplishments and future plans and goals.
2. Resource management. Our mushrooming list of physical tasks included ongoing programs for controlling fountain grass and other

alien species, establishing a permanent firebreak around the Ka'upulehu exclosure, and constructing our own water tank for irrigation.

3. Scientific research. What practically relevant information could we glean from our present outplanting project and other experiments at Ka'upulehu and elsewhere? What subjects were most important for us to learn more about to fulfill our mission of preserving and restoring Hawaiian dry forests? Can and should this science be performed at Ka'upulehu?

4. Outreach objectives. Should our primary outreach focus be on children, private landowners, or the local community? What exactly did we want to communicate to these groups, and what were the most effective techniques for doing so? Should we attempt to reach people within and beyond the other Hawaiian islands?

5. Money. Everything inevitably came back to the money circle: identifying potential sources of it, writing or presenting proposals to get it, allocating and spending it, running out of it sooner than expected, identifying potential sources for more of it, and so on.

Most of the time, it was worth it for me to log all those hours we spent collectively discussing and debating everything. Despite our different interests in and motivations for participating in the meetings and on-the-ground programs, we seemed to be united by our desire to preserve and restore what remained of Hawai'i's native dry forests. Indeed, our diversity may have been our greatest strength; none of us would have had the time, resources, expertise, connections, and desire to attempt even a small fraction of what we were now doing as a team.

We were also fortunate to have a common enemy to unite against. Unlike some other alien species that have one or more constituencies in their corner, fountain grass was hated by everyone, though not always for the same reason. The ranchers despised it because it was unpalatable to their cattle and treacherous for both man and beast to navigate. The environmental crowd wanted to eradicate it because of its devastating effect on native species and ecosystems. And members of the local community had a vested interest in controlling its growth and spread because of its proven ability to promote waves of dangerous and destructive wildfires.

Much to my pleasant surprise, all the other members of the working group seemed to be united by their common appreciation of science. Despite our painfully limited resources and continuously overflowing plate of things to do, the group consistently chose "scientific research" as one of its top priorities. Since I was the only scientist who regularly attended these meetings and performed dry forest research, I felt honored and inspired by this support. Consequently, I grew increasingly determined to design and implement a research program that would be both academically interesting and rigorous and practically valuable and relevant to our collective efforts to restore these native dry forests.

Of course, like all such groups, we had to slog our way through sometimes divisive or mundane issues such as complex budget projection spreadsheets, unintelligible government regulations, and endless administrivia. I also had to regularly educate the more practitioner-oriented people around the table about the requirements and limitations of formal science. (No, I can't answer those five complex questions with another quick and dirty experiment!) Even when we were able to focus on some seemingly straightforward component of our on-the-ground work, the process of discussing, debating, and trying to reach consensus on every little step could frustrate even the most patient members of the group. In fact, near the end of our previous meeting, which had been dominated by clashing egos and long-winded, dogmatic monologues, an exasperated elderly man who spent many hours by himself each week working in the preserve finally stood up and said, "You know, if you people would just take some of the energy and hot air you expend around this table and put it to work out in the forest, we could have cleared and planted the whole damn thing by now!"

In addition, we seemed to have to continually struggle to keep our meetings organized and productive. For instance, later that morning Andie reintroduced the important topic of how best to assess and improve our rodent control program, which largely consisted of continually replenishing the diphacinone (a blood-thinning poison) in the twenty-five plastic bait box stations scattered across the exclosure. At our previous meeting, we had discussed the fact that we had been spending increasingly more money on bait without any apparent reduction in rodent herbivory, but then we had run out of time before agreeing on what to do about this.

Several of us had promised to look into various aspects of the issue and report back to the group during today's meeting. My task was to comb through the relevant scientific literature and contact other researchers who supposedly had this same problem or were studying it. Unfortunately, as was often the case, I more or less struck out: the few "scientific" papers on rodent ecology and control in Hawai'i I had unearthed were really more anecdotal, informal reports, with little information we could really use; the scientists I contacted either never got back to me or turned out to have little data or experience that was relevant to our situation; and the more rigorous, peer-reviewed publications I found proved to be unhelpful, either because of their relatively esoteric focus or because their work had been performed in study systems radically different from ours. Finally, as also was often the case for me and most other members of the working group who struggled to keep up with the demands of our real (i.e., paid) jobs, I simply had not been able to devote as much time to my assignment as it probably warranted.

Someone who had not been at our last meeting (another perennial problem, as was the converse issue of people with key assignments from the previous meeting not attending the subsequent one) began by stating, "I think what's going on is that as we poison them, the resident rodent population is continually replenished by new rats migrating in from outside the exclosure. In fact, since it takes so long for that stuff to kill them, we're probably *increasing* the rat population by drawing them in there with all that bait! That's what somebody told me happened on his project on Maui."

"As we discussed at the last meeting," someone else interjected tersely, "the rats may not even be eating our bait—it's probably mostly going to the ants. If you look around out there, you'll see there are ant nests right next to many of the bait stations. And some of the boxes are cracked and broken, so maybe the birds or mongooses or who knows what else are getting in there as well."

"Maybe we should consider changing the diphacinone flavor we're using," one of the guys from the state's Division of Forestry and Wildlife suggested. "I've heard that the rats really seem to like the peanut butter–flavored cakes."

"You know," a representative from the US Fish and Wildlife Service cautioned, "there are strict EPA regulations on how much diphacinone we can put out there, not to mention the killing of nontarget animals. You all know how much trouble certain other people and organizations have gotten themselves into by ignoring those regulations!"

"That just demonstrates how important it is for us to work with the folks in Washington to change some of those laws," a prominent land manager said. "Look at how much success the New Zealanders have had on their islands, where they've aerially dropped diphacinone all over the place—they've virtually eradicated the rats out there, and now the birds are coming back! Why can't we do that here in Hawai'i?"

This comment led to another protracted debate between the land manager, who wanted the working group to lobby the US Environmental Protection Agency to allow his agency to perform such aerial drops, the

USFWS representative, whose agency had clashed with him on this before, and a few other people around the table with their own vested interests in this topic.

"Look," Andie finally said, "We've got to wrap this up and move on—we've got a lot of other things we need to get to today. So again, what do you all want to do about those bait stations?"

"Do we even know how many rats were out there before we started poisoning them, let alone how much damage they are really doing now?" I asked. Everyone shook their head and shrugged. "One guy I talked to told me that people generally assume that rats are responsible for all the fruit and seed predation we see on our native trees, but he thinks that alien birds are more often the real culprits. I've certainly seen lots of them eating the fruit on the sandalwood trees in the exclosure."

"I'd certainly feel more comfortable if we just stopped putting out more bait until we better understand what's going on," the USFWS representative said. "Why don't we ask the APHIS guys [the federal Animal and Plant Health Inspection Service] to come talk to our group about all this. Or better yet, maybe they'd be willing to set up and run some traplines with us so we can get some real data on what's happening out there."

"Sounds like a good idea to me," Andie said. "Are we all in agreement on this, that we'll stop replenishing the rodent bait until further notice? Okay, Bob, can you contact the APHIS guys, see if you can get one of them to come to our next meeting, and start working with them to design a Ka'upulehu rodent field study?"

"Sure," I said cheerfully, silently cursing my overcommitted self for my big mouth and inability to say no.

"Great; thanks." She walked over to the flip chart and wrote with her red "action item" marker: "Bob will invite APHIS guys to next meeting and work with them to design and implement Ka'upulehu baseline rodent field study." "Now, let's move on to our next agenda item, which is 'Outplants.' Last time, we decided to postpone our discussion about their ongoing watering regime until today. So, what say all of you?"

For once, her question was greeted with protracted silence, I think because we all knew how difficult this one was going to be. Tackling this issue was a classic example of how "simple" management practices in ecological restoration can be driven by seemingly abstract philosophical paradigms. As hard as it was for me to figure out where I personally stood on such issues, it was vastly more difficult to do this as part of a consensus-seeking working group.

After no one volunteered to start this discussion, Andie finally asked me to give an update from my most recent outplant data.

"Well," I said, leafing through my data sheets, "I only had a chance to census one of the four outplant plots this morning before the meeting, but here's where things stood as of three months ago: Overall outplant survival was roughly 80 percent—about 90 percent in the shade and 70 percent in the sun. Most of the plants had also grown more and had less damage in the shaded plots, although there was lots of variability among the plots and species within the plots."

"So I'd say on the whole they're still doing pretty well out there," Andie said. "For those of you who weren't around then, we agreed that we would provide the outplants with supplemental irrigation until they were over their initial transplant shock and had a chance to establish and develop their root systems. So are we there now? Time to cut the water off?"

"I certainly don't think so," said another member of the group who had grown up on this side of the island and probably had more planting experience than anyone else at the table. "You know, the 'dry forest' label is misleading—it used to be a lot cooler and wetter here, before people cut down all the trees. These plants used to come up in rainy periods under the shade of both the overstory forest and the understory plants. We don't have that anymore, and we don't have the luxury of waiting for a really rainy period—those have become as rare as some of the trees! You stop watering them now, and this drought keeps going," she continued, shaking her head, "a lot of those plants are going to die, and all that hard work is going to be for nothing."

"But I thought this whole outplanting was supposed to be a demonstration project to encourage private landowners and the various government agencies and nonprofits to start their own dry forest restoration programs," someone else interjected. "No one can afford to keep watering and babying their plants forever. Even if a bunch of ours die, at least we'll have learned something, and we'll be able to tell people which species are most likely to survive, and in which planting environments. Isn't that the point of this experiment, and this whole restoration program—to figure out how best to do things, so other people can take our information and start scaling up and running with it? If the future of the dry forest is these little gardening projects, we might as well just give up now and go home!"

"But I thought we were just trying to show people that it could be done," said a representative from a botanical garden. "If most of our plants die, we'll have nothing to show anyone, and everyone will continue to believe this whole ecosystem is hopeless, and no one will do anything."

We soon found ourselves immersed in yet another round of debate over our larger mission, the specific vision of the future of the Ka'upulehu Preserve, and Hawaiian dry forests in general. Should we at least try to turn the exclosure into a "real" dry forest? A scientific research site? An ever more managed and manicured garden? A museum with interpretive signage and glass display cases?

I knew that the more we intervened in the "natural" processes unfolding in our outplanting project and in the exclosure in general, the less chance I'd have to extract any meaningful scientific data out of this little experiment. Yet as much as I knew the other members of the working group valued science, I was pretty sure that if push came to shove, most of them would be unwilling to sacrifice the integrity of that forest in exchange for more scientific knowledge and accomplishments.

I had always been taught that these two goals were not mutually exclusive; on the contrary, science was supposed to provide us with both the abstract knowledge and practical tools to design and implement effective resource management programs. Indeed, one of the more politically oriented bureaucrats who frequented our meetings loved to deliver long, flowery speeches about the beautiful synergisms that would ultimately blossom among our restoration, research, and outreach programs, and thus we should never view them as competing with one another. Somehow, however, he always seemed to be elsewhere when it was time to put away the rhetoric and make the hard, non–mutually inclusive decisions such as this one.

"Well," Andie finally said, breaking in just as someone I hadn't even met was about to tell us more about his personal restoration philosophy, "we're already over time, and I know some of you have planes to catch, so we're going to have to continue this discussion once again at our next meeting. In the meantime, I guess we'll just keep going with the present watering regime?"

A few months later, Dave Lorence, Tim Flynn, and I flew back to the Big Island to quantify the effects of our now largely completed fountain grass control program by recensusing the plots Dave and Tim had established with Steve Weller the previous year.

"At least this one shouldn't take very long," I said to Tim as he handed me the metal clip of our twenty-five-meter measuring tape and embarked on his five-meter journey toward a corner of one of our fifty-three census plots. It was in a particularly rough section of the preserve, with some steep mini-canyons of 'a'ā lava and dense thickets of noxious, spiny lantana shrubs. Given Tim's work ethic and nearly obsessive meticulousness, I knew that none of this would stop him from scouring every square centimeter for potential seedlings. But because we never found much of anything growing beneath these shrubs except the occasional lantana seedling, I hoped that for once Tim might ease up just a bit.

Dave threw me the end of his measuring tape and walked in a perpendicular direction from Tim. When they both reached their five-meter marks, I fine-tuned their positions with my compass until their tapes ran exactly north–south and east–west from the plot's permanent rebar stake corner. They each then made a ninety-degree turn and kept walking until their paths crossed and their lines intersected at the ten-meter marks. After they carefully placed their tapes on the ground, we all stepped inside and got to work.

We began by estimating the "percent cover" for each of the dominant species growing within the boundaries of the plot we had just delineated. Because simply counting the number of individuals for each species is often either impossible (how many individuals are there in a clump of fountain grass?) or not informative (one large tree is probably of more biological significance than seven small herbs), ecologists often assess the relative dominance of different plants by quantifying some measure of the amount of space they occupy within a given area.

There are many ways to quantify plant percent cover in the field, including some sophisticated techniques that involve photographic paper, light meters, and expensive machinery. However, because determining the relative percent cover of different species is often more important than calculating their actual percent cover, we often rely on quick-and-dirty "eyeball" estimates. I was taught to imagine that a white sheet is lying on the bottom of the plot, and a bright light is shining down from directly overhead. In this scenario, the percent cover of a given species is equal to the percent of the sheet that would be in its shade.

To avoid biasing one another's perceptions, we always began by independently calculating our percent cover estimates.

"Okay," I said, "you guys ready?" I pulled out my clipboard. "What did you get for *lama*?"

"Fifteen percent," said Dave.

"Twenty," said Tim.

"I also got twenty," I said.

Dave walked back over to the corner of the plot that had the most *lama* coverage and looked at it again. "Okay, I guess I didn't see that all those branches were in. I'll go with twenty."

"Lantana?"

"Forty," Dave said.

"Really? I got only twenty-five." We looked at Tim.

"I was thinking fifty."

We all walked around the plot again and looked at the thicket from different vantage points. "It's covering more than half the plot," Tim said, holding out his arms at right angles to help visualize the quadrants we mentally divided the plot into to help us make our estimates.

"Yeah, but its leaves are pretty sparse, and a lot of that clump over there is dead," I argued.

"Well," Dave said, still walking around, "even still, don't you think more than a quarter of this plot is covered by lantana?"

We each looked again, trying to see it from the others' perspectives.

"Okay," I finally said. "Can you guys go with thirty-five?"

"I'll go with that," said Dave.

"Fine," Tim said.

When we finished with the percent cover data, we moved on to our "number of mature, juvenile, and seedling" data for each of the species we could count. Because this information was both less subjective and more laborious to collect, we each took responsibility for a different section of the plot. Of course, Tim immediately plunged straight down into the nastiest lantana clump and began combing through the layer of organic litter and fountain grass tufts for seedlings. "Sixteen lantana; six, no, seven fountain grass; and two . . . three . . . four *lama* seedlings!"

Dave and I scurried over to see if he was joking. But sure enough, way down near the base of the shrub's central trunk were four dark green, watermelon-like seedlings just beginning to put out their first set of true leaves. *Diospyros sandwicensis*, the dominant but up to now senescent native canopy tree of the leeward side of the Big Island, was regenerating beneath one of the most invasive shrubs in all of Hawai'i.

"I never would have imagined anything could establish down there, let alone a *lama* seedling," I finally stammered, still not quite believing what was in front of my eyes. "I was sure this lantana wasteland plot was going to be yet another big fat zero."

"Well, Bob," Tim said, smiling for once as he crawled out of the thicket, "never judge a plot by its cover."

Those *lama* seedlings were just the first of many surprises in store for us that week. By pure luck, we happened to arrive during what turned out to be the perfect storm of events for dry forest regeneration. First, the weather that spring turned out to be exceptionally cool and wet for that region of the island. Second, the fruit and seed production of the native plants at Ka'upulehu in the period preceding this benevolent spring was also exceptional (unlike the situation in temperate forests, reproduction and establishment of many tropical plants can occur during any favorable period of the year). In fact, there was such an abundance of the fleshy red *lama* fruit within the exclosure that several first-time visitors asked if it "always looked so Christmasy in here." Hannah Springer told me that she had not seen so many *lama* fruit since the 1970s. When I pointed out some *lama* seedlings to her, she asked me what they were. "Ahh," she said when I told her, smiling wistfully as she gently fondled their thick, leathery leaves, "so that's what they look like!"

The third component of this perfect regeneration storm was the success of our fountain grass control program. By that spring, we had reduced the overall abundance of fountain grass to less than 10 percent of what it had been prior to our weed whacking and herbicide treatments, and large areas that had been dense, impenetrable fountain grass stands were now effectively grass free. It was also possible that our rodent control efforts played some role, but we never really resolved the major questions surrounding that program. I did eventually run some traplines with the APHIS guys, but although we caught three different rodent species, their densities were low both inside and outside the exclosure, possibly because of the severe drought that preceded our trapping period.

As we worked our way in from the lantana shrubland that still dominated one of the back corners of the exclosure, we encountered ever more native plant regeneration. In some places, hundreds of *lama* seedlings carpeted the forest floor like moss. Nearly half of our census plots within the exclosure had at least one *lama* seedling, several had more than 50, and one, as best as we could count, had 232. Some of the endangered canopy tree species also produced viable seedlings—we even found patches of *choke keiki* ("many children") under the grandma *kauila* tree.

The most dramatic change was the new native ground cover. A native sedge and two native vine species (especially *Canavalia hawaiiensis*, a legume with beautiful purple and white flowers and big creamy, reddish-brown seeds that I loved to collect and hoard) that had barely been present the previous year now blanketed sections of the exclosure that had been fountain grass monocultures. It was especially gratifying to see the vines, whose previous distribution had been largely restricted to tree trunks and branches, grow or disperse out into the open and, in some cases, even smother an isolated fountain grass clump or juvenile lantana shrub. Several other herbaceous and shrubby native species increased their distribution and abundance in a similarly dramatic fashion.

The combination of all this multispecies, multilayer native regeneration made a few choice sections of the preserve actually look like a normal, healthy forest. But as I discovered while showing off this spectacle, it was almost impossible for the average ecologically uninformed and inexperienced visitor to appreciate the significance of this phenomenon. I did not fully grasp it myself until several prominent botanists independently told me they had never seen anything like this occur in any other low-elevation, accessible area in all their years of exploring the Hawaiian archipelago.

However, all was not utopia at Ka'upulehu. By the end of that week, I realized that the ecological story unfolding in front of us was considerably more complicated than I had first appreciated (or, as Tim put it, "The plots keep thickening"). As it turned out, the natives were not the only species able to reap the benefits of our efforts at fountain grass control. In addition to the explosion of several previously minor weeds, we found sixteen new, potentially noxious alien plant species that had never before been seen within the exclosure. Presumably, much as we had hypothesized in our National Science Foundation grant application, prior to our war on fountain grass the presence of this dominant alien species had effectively suppressed the colonization, establishment, and spread of both native and alien plants. Indeed, fountain grass itself appeared to capitalize on its own eradication: whereas we had not found a single fountain grass seedling in our previous year's census, this time we found them virtually everywhere.

Perhaps the biggest surprise came when we climbed over the fence and into the surrounding cattle- and goat-infested fountain grass. When Dave, Tim, and Steve had first established those five-by-five-meter monitoring plots, they wisely decided to also sample the vegetation in the unfenced regions adjacent to the Ka'upulehu Preserve. Since no one had ever killed fountain grass and rodents out there or probably ever would, these data could serve as an unmanipulated reference point for what happened inside the exclosure.

Unfortunately, this adjacent area was floristically quite different from the area inside the exclosure. Whereas the preserve still contained a relatively diverse and abundant native flora well before we started our restoration efforts, many decades of domesticated and feral ungulate activity had apparently converted much of the unfenced adjacent area into an almost pure monoculture of fountain grass. Because of these differences, it was a stretch to consider the adjacent area as a formal experimental control for what we observed within the preserve in response to our restoration efforts, but, once again, it was simply the best we could do. While one could (and some did) argue that we should have randomly divided up the exclosure into treated and untreated regions, this would have been an impractical, costly, and ethically questionable strategy to pursue in such an endangered and tiny forest fragment.

When they sampled the vegetation outside the exclosure, they had not established permanent five-by-five-meter plots because Hualalai Ranch's cattle might well have trampled their stakes, and because it had been enough of a struggle to obtain the ranch's permission to be out there in the first place — the last thing ranchers in this land of endangered species want is a bunch of highly trained botanists snooping around their property! Consequently, they simply sampled a series of five-by-five-meter plots along transects that ran parallel to the preserve's fences. Because that area was so homogenous, they decided to sample only about half as many plots as they had within the exclosure (because there is an inverse relationship between ecological variability and statistical power, we generally don't have to sample uniform areas as intensely as diverse ones).

When I first went over the fence with Dave and Tim to sample those transects again, the whole affair struck me as somewhat absurd. There we were, three grown men struggling to make our way through an unbelievably dense fountain grass jungle (the wet spring had apparently fueled an unprecedented amount of growth in this species as well) just so we could say "Yup, good scientists that we are, we dutifully collected our comparative quantitative data, and guess what: it's still just a fountain grass wasteland out there!"

The grass was so tall that we had to hold our tape measures over our heads to run the line transects, and so lush that we had to struggle to pull them down low enough that we could see the boundaries of each plot. Yet it really didn't make much difference where the boundaries were because almost the whole area traversed by our transect lines was just one treeless, amorphous mass of fountain grass broken only by a few relatively barren 'a'ā lava outcrops and clumps of lantana. After a few such plots, even Tim became a little less diligent about clawing his way to the ground just to confirm that there was nothing other than fountain grass down there. I was tired, hot, and scratched up enough to start considering whether anyone would ever know if we just agreed to copy and paste in the census data from last year's transects and hit the beach.

But then, near the end of the first transect, we came to the one region outside the preserve with a significant *lama* overstory. We had looked at this area before from within the exclosure and wondered why it still contained so many native trees. As best we could tell from its topography, it may have once been a wide lava tube that had collapsed and left steep, rough banks on both sides of its sunken interior. Had these geographic features initially prevented, or at least deterred, the resident ungulates from destroying the trees that eventually colonized and grew out of this former tube? Or did this long tubelike depression create a cooler, moister microenvironment for the trees? Maybe it was both of these factors; maybe it was neither.

As luck would have it, our first census plot within this area fell directly beneath two *lama* trees. After we recorded our percent cover data for the overhanging canopy (75 percent) and the fountain grass understory (90 percent; the remaining 10 percent was bare lava), I bent over to do my cursory search for seedlings. To my utter disbelief, as soon as I parted the grass enough to see the ground, I saw *lama* seedlings everywhere. "Hey!" I yelled to Dave and Tim. "You're not going to believe this!" But both of them were already on hands and knees, marveling at their own dense seedling patches. We painstakingly counted 195 *lama* seedlings within that plot, and the next one that fell beneath *lama* trees had 183.

I was dumbfounded; how any plant could get enough light or water or nutrients within all that thick fountain grass was beyond me. What did this mean?

My mind raced with new hypotheses and experiments to explain and test some of what we had seen and learned that week. Maybe that local resident was right—maybe, given the present hot and dry environment of this degraded ecosystem, native dry forest species need a protective, densely vegetated niche in which to germinate and establish. Given the inability of the native dry forest canopy trees to reproduce on their own here and throughout the other Hawaiian islands, some kind of intervention was clearly necessary. But rather than trying to eradicate all the noxious weeds, maybe we could pull off the proverbial "lemons into lemonade" trick by manipulating alien species such as fountain grass and lantana in such a way that they *facilitated* rather than suppressed the regeneration of the native species. I knew that the establishment of some plants in the desert ecosystems of the American mainland often occurred within the relatively benevolent microenvironments created by other so-called nurse species; maybe we could figure out a way to make that same kind of ecological process work here.

Alternatively, or perhaps in addition to the alien nurse plant approach, we could try to manipulate assemblages of relatively fast-growing native species so that they created favorable niches for eventual establishment of the endangered, slower-growing canopy tree flora. Maybe we should try sowing seeds and planting seedlings of canopy trees under native shrubs covered over by native vines. Compared with the time and effort necessary to collect, propagate, and outplant the canopy tree species, creating something like a *Canavalia*-encrusted shrubland would be a breeze.

We could also complement those kinds of experiments with descriptive observational studies in which we monitored the fate of all the native and alien seedlings that came up on their own. Which of those thousands of *lama* seedlings would ultimately establish into saplings without any help from us? Would any survive outside the exclosure, or would the cattle and goats eat and trample every last one? If we protected them from the ungulates, could they grow up through all that fountain grass? What if we thinned, or watered, or fertilized the grass . . . ?

While my scientific brain hummed along and continued to formulate what I hoped would be rigorous and eloquent ways to poke and prod this ecosystem into revealing its secrets, my conservationist's heart both rejoiced at what was happening and feared it might be short-lived. Had we really demonstrated that with sufficient effort there was hope for even this brutalized ecosystem, or was it pure delusion to think that we had turned the tide here? If all we did was just experiment and monitor, rather than intervene and manage, would fountain grass or some of those new weeds come roaring back? What if all those native tree seedlings died, and this ultimately proved to be the very last episode of "natural" recruitment for some of those species before they became extinct in the wild?

Dave and I unpacked our lunches, sat down, and leaned against the sturdy trunk of one of the larger *lama* trees in the exclosure. I was happy to be in the shade and even happier that for once we had completed our fieldwork ahead of schedule. As usual, I had been worried that we weren't going to finish our census before we had to return to Kaua'i. While I could have stayed longer, I knew that Dave and Tim had to get back, and I couldn't do the work without them.

When we had finished our lunch, and Tim had finished pulverizing a patch of invasive prickly pear cactus (*Opuntia ficus-indica*) with a stray piece of rebar and stomping down a nearby lantana patch with his boots, the three of us slipped on our packs and headed back down to our rental car. I was sad to leave the forest and all the exciting things that were happening in it, but I knew it would be good to get away and spend some time in the office processing everything and carefully thinking through what my next steps should be. I was also looking more than a little forward to soaking my week's worth of wounds in the ocean and finally spending some real time on the beach.

"Whoa, where'd that come from?" Dave asked, jolting me back from my little daydream. He pointed to a shrubby plant that had popped up just inside the fence that separated the forest from the highway. It was about four feet tall, with branching spikes at the top loaded with tiny dry fruits about to burst.

"Careful," Tim warned when I started handling some of its dry capsules. "Don't scatter any seeds." Though the plant looked vaguely familiar, I had no idea what it was.

"*Buddleia asiatica*—butterfly bush," Dave said. "Add it to your list of new weeds, Bob—we've never seen it in here before, but it's a nasty one, and it can take off in a heartbeat."

Tim pulled a garbage bag out of his pack, draped it over the shrub, and then bent down, ripped the plant out of the ground, and deftly flipped the whole thing over so that all the seeds and branches remained within the bag.

"Are you sure it would have spread in here if we let it go?" I wondered out loud. "Maybe we've tipped the scales enough so that the natives could have at least held it in check?"

"Trust me," Tim said earnestly, tying up the bag, "you've got to nip weeds like this in the *Buddleia*."

Chapter 4

Writing It Up: The Art and Importance of Science Papers

One week after Dave Lorence, Tim Flynn, and I completed our fieldwork at the Ka'upulehu Dry Forest Preserve, I finished entering the last of our census data into the computer. Before leaving the office, I saved it on the hard drive, backed up the file on two disks (one of which I'd keep at home in case the office burned down), e-mailed copies to myself and Steve Weller, and stowed the original field data sheets, along with a printout of the computer spreadsheet—162 rows by 96 columns—in a carefully labeled folder. I had learned the hard way in graduate school that it was always wise to be meticulous and paranoid with scientific data and computers.

I had also been taught that no matter how hard we work, how important our results, or how brilliant our analyses, our job is not done until we publish our science in a reputable, peer-reviewed journal. This is because such papers are the gold standard of science. In much the same way that judges allow some forms of evidence into their courtrooms and reject others, scientists consider the data and conclusions in this formal literature to be credible, "permissible evidence," and tend to ignore or discount forms of knowledge that have not been subjected to the rigors of the peer-review process.

When I started educating myself about Hawaiian conservation biology and restoration ecology, I found a plethora of personal opinions and popular writings but depressingly little real scientific literature. Consequently, I often had to conclude that we didn't really know much at all about many of the most basic and fundamental topics in these disciplines. Thus, I wanted to do my small part to help remedy this situation by formally documenting our work at Ka'upulehu and disseminating it in a format that other scientists and practitioners within and beyond the Hawaiian islands could utilize and build on. On a more selfish level, I also knew that the more peer-reviewed publications I produced, the greater would be my chances of eventually landing a job in the hypercompetitive worlds of professional ecology and conservation.

Although entering and error-checking lots of field data is tedious and time consuming, this is really the easy part for me. The hard part is all the deep thinking necessary to make those numbers tell a rigorous and interesting story that fits into the standard "Introduction, Methods, Results, Discussion" mold of formal science papers.

Writing these kinds of papers is far more difficult and laborious than most people realize. It can take weeks or even months of reading, discussions with colleagues, and exploratory data analyses just to come up with a suitable framework in which to put the pieces of the typically complex scientific puzzles we must conceptualize and build. Deciding which pieces to keep, how best to fit them together, and how to connect the completed puzzle to an appropriate larger story are also often agonizingly slow and frustrating tasks.

Even if some practitioners were able to design and perform projects suitable for peer-reviewed academic journals, very few would have the necessary background, time, and incentive to convert these projects into formal scientific papers. Thus, in much the same way that scientists who have never done any applied, on-the-ground restoration tend to underestimate the time and effort such work requires, few practitioners or nonscientists grasp how much difficult work lies behind even seemingly straightforward science papers. This mutual lack of understanding and appreciation of one another's work is an important component of the science-practice gap in restoration ecology and other related disciplines.

I knew that transforming the various pieces of our Ka'upulehu dry forest puzzle into a coherent scientific paper was going to be particularly difficult because our study had not been explicitly designed as a formal research project. Moreover, descriptive investigations such as this are generally much harder than experimental studies to fit into the reductionist frameworks, involving specific questions and hypothesis testing, that peer-reviewed scientific journals expect. Even if our experiments don't turn out the way we predicted, at least we have an a priori rationale for our specific treatments: "I provided three levels of fertilizer to test whether the relationship between increased nutrient availability and maternal genotype would . . ." or "I excluded all birds from each flowering tree to investigate whether insect pollination could . . ." But merely passively observing nature to see what happens can result in some pretty murky science, both because it is easier to do without first formulating a clear reason for doing so and because it is more difficult to know exactly why things turned out the way they did.

If we had, say, killed fountain grass in or sowed *lama* seeds into half of our five-by-five-meter census plots at Ka'upulehu, we could then have confidently concluded that any consistent differences we observed between the experimental and control plots were caused by our treatments. The quantitative analysis of these data would also have been relatively straightforward—we'd simply arrange each measured variable (number of *lama* seedlings, percent cover of native herbaceous species, etc.) into "treatment" and "control" columns and use standard statistical procedures to test which variables differed significantly between these two groups. But in our case, as so often happens when science and applied projects are intertwined, things weren't quite so neat and tidy.

To be sure, I knew this study had some major strengths—that's why I decided to devote a substantial amount of my increasingly limited time to spearhead our efforts to publish it in a first-rate journal. Unlike the situation with many such projects, which tend to be heavy on "resource management" and light on "scientific design and data collection," in our case Steve, Dave, and Tim had the necessary training, skill, and patience to design and implement a rigorous data collection scheme. The fact that our results suggested there might be some hope for this globally endangered, degraded, and understudied ecosystem also greatly increased both its academic and practical value. At the same time, however, because the top journals were receiving ever more submissions each year, they had to be increasingly selective. Thus, whether our manuscript would be accepted or rejected would largely depend on the extent to which we were able to "sell" our story to the journal's necessarily choosy editors and outside reviewers.

But what exactly was our story? What were our specific questions and hypotheses?

I knew that writing "We carefully recorded the floristic changes that occurred over a two-year period in a series of plots established within and outside of the Ka'upulehu Preserve" wasn't going to get us very far. And even though the fieldwork had been exceptionally well designed and executed, the conceptual model underlying this work was really not much different from the standard "Let's set up a series of baseline monitoring plots to document what happens" approach typically employed by less formally educated practitioners. Clearly, we would have to come up with something better, even though I knew from yet more hard experience in graduate school how difficult it can be to formulate good a priori research questions and hypotheses for studies after they have been completed.

We finally decided that the best strategy would be to superimpose our resource management treatments onto our census plot data. As luck would have it, even though the weed whacking and herbicide treatments were initiated during the winter of 1995, my colleagues informed me that these efforts had not yet significantly affected the preserve's fountain grass population by the time of their first censuses in the spring of 1996. Also fortuitously, the rodent-poisoning program had not begun in earnest until the winter of 1996. Thus, even though it had not been explicitly designed this way, we could legitimately consider the 1996 census as the "before alien species control" data and our just-completed 1997 census as the "after alien species control" data. We could also use the two years' worth of census data collected in the untreated adjacent area outside the exclosure as at least a pseudocontrol for the differences between these two years. That is, if the changes we observed between the 1996 and 1997 exclosure census data were purely the result of, say, the greater amount of rain in 1997, then we would also expect to see similar differences in our untreated plots outside the exclosure.

Although it had also not been explicitly envisioned this way at the time, we now realized that the 1996 census data, along with some of Dave and Tim's other previous work (particularly their exhaustive survey of the entire preserve's flora and their quantitative measurements of all its woody species in 1995), could be legitimately framed as an investigation of the question of whether more than forty years of ungulate exclusion was sufficient by itself to preserve the flora of a tropical dry forest. As far as we knew, no other dry forest in Hawai'i or elsewhere had ever been protected from ungulates for this long.

After a few more rounds of exploratory data analyses, reading, discussion, false starts, and dead ends, I felt ready to take a stab at translating our nebulous ideas into more precise and formal academic writing. As with all my previous scientific papers, I found that once I finally boiled it all down into a few deceptively simple statements, in retrospect it seemed almost obvious. In the manuscript we eventually submitted (titled "Effects of Long-Term Ungulate Exclusion and Recent Alien Species Control on the Preservation and Restoration of a Hawaiian Tropical Dry Forest"), the wording of that crucial concluding paragraph of the Introduction, in which authors are expected to state the specific motivation for their studies, turned out like this:

Our study compared vegetation in the Kaupulehu Dry Forest Preserve in the North Kona district of the island of Hawaii, a fenced area established over 40 years ago, to an adjacent unfenced area. After initial censusing of this exclosure in 1995 confirmed that virtually no regeneration of native canopy trees had occurred despite the long-term absence of ungulate grazing, we initiated an aggressive management program designed to control the alien fountain grass and rodent populations within the Kaupulehu preserve (hereafter "preserve"). To assess the potential effects of both long-term ungulate exclusion and the more recent alien species control program, we compared the flora inside the preserve with the flora of the unprotected surrounding adjacent area (hereafter "adjacent area") before and after the control of the fountain grass and rodent populations.

With this preamble in place, I was ready to tackle what for me is always phase two of writing a science paper: data analysis. Because there is an almost infinite number of ways to approach and massage scientific data sets, I knew how easy it could be to get stuck in the dreaded "data analysis whirlpool." This was one lesson I had managed to learn vicariously. One day in graduate school while I was working in the computer pod, another student came in with a stack of statistics books and programming manuals. She explained that she had just finished entering all her dissertation data and had come in to use our mainframe statistics program to analyze it.

"How long do you think that's going to take you?" I asked.

"Oh, no more than a day or two," she replied matter-of-factly. "I know what I'm doing, and this program is really fast and easy." She booted up the machine and got right to work. "I can't wait to see how this is going to turn out," she added excitedly. "The suspense has been killing me!"

I found her there virtually every time I went to the computer pod that summer. And she did know what she was doing—she was an exceptionally bright and hardworking student. By the end of the summer, she had produced an impressively thick ream of graphs, tables, and complex statistical output. Yet, because I knew she still had not written a single word of her dissertation, I asked another, more experienced graduate student in her lab what was going on.

"She never really developed any specific questions for her work before she went into the field," he replied immediately. "So now she's trying to retrospectively figure it out by doing all that data manipulation and analysis and modeling. That's a long, hard, *back-assward* way to do it!"

Ever since that conversation, I've always tried to have at least a ballpark sense of the specific questions I am trying to answer before I start slicing and dicing the data. Of course, that doesn't mean that data analysis is quick and easy for me, or that my initial research questions and rationale don't often get tweaked by my actual results. But so far, at least, this strategy has enabled me to swim out of rather than drown in the data whirlpools that seem to catch many other scientists.

I began by making a list of specific questions suggested by the research framework articulated in the concluding paragraph of the Introduction. Next, I tried to figure out what general approach to use in answering each of these questions. Once I had a reasonably clear conceptual vision, I tackled what for me has always been the much more difficult task of applying these ideas to my typically problematic data. (Somehow, practically every data set I've ever analyzed has been fundamentally different from and more complicated than the examples in my statistics classes and textbooks.)

For example, one of the first specific questions I wrote was "How did our alien species control programs affect plant regeneration?" While it took me only a few minutes to think up this question, figuring out how to answer it took far longer. Because there were essentially no saplings of any of the native tree species at Ka'upulehu, and we couldn't determine the ages of the shrubs (an individual shrub could be small because it was young or because it was old but slow growing), I ultimately decided to answer this question solely via an analysis of our seedling data. But directly comparing the number of seedlings in the preserve with the number in the unfenced adjacent area was not a straightforward problem. First, there were fifty-three preserve plots but only twenty-six adjacent area plots. Second, because the seedling data were extremely patchy, with none in most plots and tens or hundreds in a few others, I couldn't use standard statistics to analyze them because these kinds of parametric tests require data sets with more even, "normal" distributions. And third, combining and comparing the seedling counts for each of the different species was a bit tricky. Should I compare the total number of seedlings between these two areas, and across the three years of the study, or the total number of native versus alien seedlings? Would it be more informative to break the seedling data into different life history stages (e.g., trees versus shrubs versus herbs) or to do individual species-by-species comparisons?

Resolving these kinds of questions required many more hours of exploratory investigation. While doing this work, I always have to struggle to make my final decisions as objectively as possible. I know that even when

I think I am being a fair and impartial scientist, it is all too easy to unconsciously "fish" the data until I find whatever best supports my preconceived ideas or produces the most dramatic and interesting results. In this case, I knew that I "wanted" the data to show that there were significantly more native seedlings in general, and canopy tree seedlings in particular, in the preserve following our alien species control programs. Thus, without actually falsifying any information, I could have arranged and tested our seedling data in dozens of different ways and then selected the one or two that best demonstrated my desired patterns.

The temptation to rationalize such fishing expeditions can be overwhelming, especially when doing so may substantially increase one's chances of getting a paper published, a proposal funded, or a promotion granted. Nevertheless, perhaps because of an intrinsic love of and commitment to good science, the great majority of formally trained professional researchers I know strive to be as objective and rigorous as possible. Journal editors and the expert independent reviewers they rely on also tend to be highly skilled at detecting and correcting potentially biased methodologies, analyses, and interpretations that sometimes creep into our work despite our best intentions.

For this study, I finally decided to compare the *proportion* of plots in the preserve and the adjacent area with and without seedlings of the most common species in each census year. This strategy solved both the problem of there being more plots in the preserve (proportional comparisons are unaffected by total plot number) and the patchiness of the seedling data itself (since I used a presence-or-absence approach, plots with one seedling of a given species had the same weight as plots with one hundred). Excluding all the species with seedlings in only one or a few plots also enabled me to focus on a relatively small group of key species and produce an uncrowded graph that was easy to interpret. Finally, by using a relatively simple nonparametric procedure called a chi-square test, I could calculate the probability that the proportion of plots with seedlings of a given species was significantly different in the preserve and in the adjacent area.

Despite my efforts to employ a systematic, narrowly focused approach throughout the data analysis phase, I always end up with far too much output for a single scientific paper. This is a problem because all journals strive to make their articles as short as possible, and since data tables and figures require a relatively large amount of space, authors are generally allowed to include only a small handful of them. Consequently, in much the same way that I struggle to perform each individual analysis in an objective and nonbiased manner, I often agonize over which sets of analyses to exclude from the final paper. However, once I make these decisions, I force myself to put all the unused mountains of data and statistical output away and try not to even think about them again until the manuscript is submitted.

The next step in my personal "How to Write a Science Paper" program is to tackle the Results section. The purpose of this "just the facts" section is simply to present and explain one's findings as clearly and concisely as possible. Although it can be tedious, I find this to be the easiest section to write, perhaps because by this point I have already done the relevant heavy thinking and struggling in the data analysis phase, and thus all I have to do here is talk the reader through each of my previously completed graphs and tables. Yet I know that this is often one of the harder sections for other scientists, perhaps because many of them do their Results writing and data analysis simultaneously. (I've tried to follow some of their models and have managed to get a few to try my program, but the outcome has generally been disappointing in both directions. This result highlights the underappreciated fact that there usually is a great deal of creativity and subjectivity lurking beneath the surface of even the most rigorous peer-reviewed science.)

With the data analysis, the Results section, and the concluding paragraph of the Introduction complete, I am ready to take on the Discussion. This section usually includes an attempt to directly answer the questions posed in the Introduction, an interpretation of the significance of these answers, and a discussion of the larger implications of the study as a whole. The Discussion is always the hardest section for me to write because it is relatively free-form and open-ended. I also find it difficult to strike the right balance and avoid being too conservative or too liberal. It is easy to be too cautious—make only narrow, meek arguments; hedge and qualify everything—and wind up saying nothing very interesting or new. On the other hand, it is equally easy to be too speculative and make sweeping, bold arguments that are not sufficiently supported by or relevant to the study's actual results. Most editors and reviewers quite rightly have a low tolerance for overly liberal Discussions.

Once I complete my initial, relatively narrow discussion of the answers to the study's specific questions, I always spend a lot of time pondering where to go next. The absolutely hardest section for me to write is the end of the Discussion, in which I must state what I believe are the study's most important larger implications and take-home messages. By this point, I often have the urge to write "*I don't know what it all means!*" and move on to something new that promises to be more fun and less complicated.

In this paper, I decided to end the Discussion by placing our results within the larger framework of ungulate exclusion studies. After reviewing the relevant literature, I wrote that "the extent to which native plant communities recover following [ungulate] exclusion is often inversely related to the amount of degradation already experienced by the area prior to fencing," and thus "the current highly degraded state of these [native Hawaiian dry forest] communities suggests there is little hope that they can eventually recover without intensive management and reintroduction of native species." In other words, my argument was that this study and others suggest that ungulate exclusion is a necessary but by no means sufficient step in the preservation and restoration of trashed ecosystems like this one.

Okay . . . now what? After much agonizing, long discussions with my colleagues, and prolonged bouts of skillful procrastination, I found myself repeatedly returning to my mental images of Tim bludgeoning that patch of prickly pear cactus and ripping out and removing that solitary butterfly bush. What would happen if he, and collectively we, hereafter just left the Ka'upulehu Preserve alone?

This question seemed to tie together all my disparate scientific and academic interests and conservation and management concerns. As an academic scientist, I knew that carefully monitoring what happened within the Ka'upulehu exclosure in the absence of any further intervention would very likely yield an intellectual gold mine. For example, because there was a rich body of theoretical literature that attempted to model and predict how new species invasions might progress over space and time, we could elegantly test some of these theories by mapping the spread of each new weed that colonized the preserve. We could also zoom out and investigate whether (and if so, how) our relatively brief interventions altered the trajectory of this ecosystem as a whole. For instance, as discussed in our National Science Foundation grant application, would the removal of an existing dominant alien species result in the establishment of a new dominant alien species? Could one or more formerly suppressed native species establish self-sustaining populations? What effect might a new, post–fountain grass ecological equilibrium have on abiotic processes such as nutrient and hydrologic cycling?

However, although this approach could also yield data and insights that would help inform and guide our efforts, and those of others, to control alien species and restore native ecosystems, the thought of implementing it made the conservationist in me a bit queasy. Was the prospect of excellent science that might be of practical value worth the risk of potentially severe and irreversible ecological damage to such an endangered and laboriously restored forest? How would the other members of the North

Kona Dryland Forest Working Group feel if we passively stood by and watched the exclosure become inundated once again by fountain grass or other noxious weeds? Or could we find some middle ground, in which we would rigorously investigate at least some of these academic questions while minimizing the risk of further ecological degradation?

Since I had no answers, I finally decided to pose some of these questions, briefly discuss their implications, and hope I didn't sound too wishy-washy.

At last, the moment of truth had come: time to take a crack at the paper's final paragraph. In my book, a good Discussion section should build, crescendo-like, to some kind of pithy, big-bang finale. Adding to this pressure was the seldom admitted truth that for all but the few papers directly related to our particular subdisciplines and specific research programs, many of us harried scientists will look at a journal article's title, skim its Abstract and Introduction, skip the Methods and Results entirely, skim the Discussion, and then slow down and carefully read the concluding paragraph word for word.

I stared out my office window at the large, empty ocean and imagined people in Kansas or South Africa with this paper in their hands. What did I want them to take away from this study of a tiny, desperate forest on a volcanic rock in the middle of nowhere? Something clicked, and for once I had a real answer to one of my questions: "Don't ignore what's happening out here! Yes, Hawai'i is a strange and unique place, but it's not irrelevant—an increasing portion of the natural world is heading in this direction, and fast. If you think the places you love in your neck of the woods will be there forever, think again! Act now, before they become so degraded that it takes heroic efforts to save them. But don't give up without a fight on the places that have already been trashed, and don't give up on us out here in Hawai'i. Nature can be surprisingly resilient when given half a chance, and even remnant little postage stamps like Ka'upulehu have value."

Spitting that out was surprisingly easy, though translating it and some of my colleagues' additional input into more formal journal language took much longer. Here is how it ultimately came out:

> Hawaii's unique location and biogeographic history may be largely responsible for the high proportion of alien species that have become established in these islands. Nevertheless, the ever-increasing global spread of exotic species (e.g., Drake et al. 1989; Devine 1998) suggests that some mainland ecosystems may eventually ex-

perience similar magnitudes of alien species invasions. In Hawaii it has become increasingly clear that without active management, remnant stands of native vegetation will suffer further deterioration, ranging from gradual loss of native species to complete ecosystem destruction. Although the costs of aggressive intervention may be high initially, over time these costs often decrease dramatically as major threats are controlled or at least mitigated. Once preserved, these remnant communities may then be utilized for biological and cultural education, as habitat for rare species reintroductions, and as a model and propagule source for future, larger-scale ecosystem restoration.

I learned in graduate school that science papers should have an over-arching hourglass structure: start broad, narrow to the specifics, broaden back out. Yet, as I have tried to illustrate here, I have found that the only way I can accomplish this is to do it from the inside out (start with specifics, broaden back out, then start broad). For me, a major advantage of this approach is that it makes the beginning of the paper (which many of my colleagues dread the most) relatively easy to write. By the time I get there, rather than staring helplessly at a blank screen, I already know what the results and their subsequent interpretations and take-home messages are, so I know what the broader framework is. Similarly, since I have already written the end of the Introduction, I know exactly where this section needs to go to connect to that final, narrowing paragraph containing the paper's specific research questions. Last and perhaps most important, by this point I'm almost done with the damn thing!

So I sat down with all the relevant literature I planned to cite and discuss and started smugly composing my Grand Introduction. Yet after three long days of dreary labor, all I had to show for my efforts was a bunch of discombobulated, tortured prose. How did I con myself into believing that this time it really was going to go quickly? Once again, I proved that even when I know where I want to start, what I want to say, and where I want to end, this kind of writing is *never* quick nor easy. (I have known a few scientists who seem to find this work relatively easy. But none of us like these people very much and generally wish them ill.) Perhaps this is why the ultimate fate of so many otherwise excellent studies is to be buried alive as partially completed manuscripts in someone's filing cabinet.

After I finally whipped the Introduction into shape, all that remained was undemanding grunt work. Writing the Methods section was largely just a matter of hunting down and assembling all the miscellaneous facts and documentation pertaining to our study site, fieldwork, and data analyses. Then came the Literature Cited section, which is really just another form of data entry, and then, at last, the Abstract, which often does go relatively quickly because its skeleton can be pasted in verbatim from the paper's other sections, and editing is generally far easier than composing.

With this now complete draft of the manuscript in front of me, I sat back and for the first time read it straight through from beginning to end. In addition to tightening and polishing the paper's writing and logic, I pretended I was an outside reviewer whose job was to assess this study and decide whether or not it warranted publication. While it is impossible to objectively evaluate one's own work, it is often surprisingly easy to criticize it. Indeed, I have to struggle not to be overly critical of my own work, as I am always painfully aware of its many flaws and limitations.

This study certainly was no exception. As I read the paper, I kept a running list of all the questions and criticisms I would have if I were one of its outside reviewers. This list wound up containing some things I had been worried about all along, as well as a few issues I hadn't thought of before. For example:

- You have no information about what this area was like prior to the time that the Ka'upulehu Preserve was fenced. Thus, at least some of the differences you observed inside and outside this exclosure could be due to preexisting differences between these two areas, rather than the effects of ungulate exclusion per se.
- Because the alien species control treatments were performed in a different area from the adjacent "control" area, you have a split-plot design with no replication at the treatment level. Thus, your entire study is pseudoreplicated—sampling only one preserve and one adjacent area is statistically invalid and meaningless!
- How can you claim that the native regeneration you observed within the exclosure in 1997 was the result of your alien species control efforts? I don't buy your argument that if this had been caused by the wet spring, you would have seen lots of native seedlings in the adjacent area as well. Your data clearly show that there were virtually no natives out there, so unless you sowed native seeds outside the fence (why didn't you do that, by the way?), how could they possibly establish in the adjacent area on their own? Moreover, you did find lots of seedlings in the adjacent area of the one native species (*lama*) that is still out there and apparently producing viable fruit.
- Although interesting, your study is still quite preliminary. It's hard to

know what to make of it with only two years' worth of data. I would thus encourage you to collect another year or two of census data and track the fate of all those *lama* seedlings and new weeds. Depending on what you find, perhaps inclusion of these data will then make this study worthy of publication in this journal.

If only it were as easy to address these kinds of criticisms as it was to levy them! Yet I knew that these were fair and reasonable concerns. For most of the criticisms I came up with, about all I could do was acknowledge that we were aware of these issues (without, I hoped, bringing up something that readers wouldn't have realized on their own), explain why we did what we did, try to be sufficiently careful with our interpretations, and pray this was enough to placate our actual reviewers and editor.

When I reached that familiar saturation point, where I absolutely couldn't bear to even think about this study anymore, I knew it was time to send it to my coauthors for one final round of their input, and to a few trusted colleagues outside Hawai'i for a round of "friendly review" (i.e., do what I had just done and critique the paper as harshly as possible). Although it can be unpleasant, it's always better to get criticisms from one's friends, when there is still time to address them, than it is to get them from the journal's editor and reviewers, who may or may not give you a chance to respond and resubmit the manuscript.

After one final week of correcting, tweaking, and a bit of backpedaling, at last the time had come to write a gracious cover letter and drop the manuscript in the mail. We had decided at the outset to submit this study to *Conservation Biology*, a widely read, well-respected, and highly selective journal that tries to reach both the scientific and practitioner communities.

I sealed the envelope and felt a tinge of pride at its hefty weight (the final version was forty-three pages long). I stepped out of my office and blinked in the midday sun like someone who has just emerged from a long stay in a dark cave. The wind was rattling the eucalyptus trees around the NTBG's administration building and blowing whitecaps across the ocean down below. Although I felt as if I had been immersed in the world of Hawaiian restoration ecology and conservation biology for most of my adult life, I realized with a jolt that I had landed on this island for the first time almost exactly one year ago.

With a mixture of excitement, insecurity, exhaustion, and relief, I dropped my precious package in the mailroom. Would I have trouble even remembering what this paper was about by tomorrow morning, as had been the case with all my previous manuscripts, or would things be different now? Either way, I knew that it was really, really, *really* time to head to the beach.

About six months later, the editor of *Conservation Biology* informed me that our manuscript had been rejected. However, it was an extremely kind and thoughtful rejection (I've had plenty that weren't), and even better, the editor encouraged us to resubmit our work after addressing his and the two anonymous reviewers' concerns. What turned out to be that study's Achilles' heel—that it was too preliminary and needed at least another year's worth of data—was the one major fault I had anticipated but not explicitly addressed. This was because none of my reasons for submitting the paper with this fault seemed scientifically valid: I was in Hawai'i on a temporary postdoctoral fellowship and might be long gone next spring, I thought it would be difficult to collect another year's worth of meaningful data because more and more people were coming into the tiny exclosure and altering its flora, and I felt that Hawaiian dry forests desperately needed as much immediate attention and support as possible. But the editor and reviewers appeared to have appreciated the study's strengths and the ecosystem's urgent plight, and they thus implied that if we collected data for one more year and addressed a few other, relatively minor concerns, there was a good chance they would accept and publish a revised version of this manuscript.

As it turned out, shortly after submitting that paper, on a whim I applied for and got a permanent job as a restoration ecologist with the USDA Forest Service, based in the town of Hilo, on the eastern side of the Big Island. Because the National Science Foundation also funded our revised grant proposal, much to my pleasant surprise, I ended up having the necessary economic, logistic, and institutional support to continue and expand my dry forest research program at Ka'upulehu.

Chapter 5

Scaling Up: Micro to Macro Science and Practice

In collaboration with the North Kona Dryland Forest Working Group, over the next several years an expanding network of colleagues and I embarked on numerous scientific research and on-the-ground restoration programs within the six-acre Ka'upulehu Dry Forest Preserve. Although each project had its own unique goals and methodologies, our collective efforts zeroed in on what we believed was the crucial overarching restoration question for this ecosystem: what is the best way to simultaneously control fountain grass and facilitate native plant establishment at ever larger spatial scales?

Piece by often painstaking piece, we tried to learn enough about this ecosystem to be able to put some form of the original puzzle back together in a semi-informed and intelligent manner. To accomplish this goal, we employed a mixture of carefully planned, formal scientific investigations and more opportunistic, ad hoc experiments and observations that often capitalized on a series of fortuitous events.

For example, because the next winter and spring at Ka'upulehu were exceptionally dry, there were markedly fewer seedlings than there had been in the previous wet spring, when we found all those naturally regenerating seedlings. Consequently, even though there had in fact been lots of subsequent weeding throughout the preserve (including the efforts of one heroic teenager whose vitriol for lantana would have made Tim Flynn proud), we were still able to collect another year's worth of meaningful census data because that prolonged drought had halted the establishment and spread of the great majority of both native and alien plant species. Fortunately, we had also previously decided to track the fate of all those naturally recruiting *lama* seedlings by marking a series of seedling patches in both the preserve and the unfenced adjacent area. To separate the potential effects of fountain grass competition from the effects of cattle and goat grazing and trampling, we had placed wire mesh ungulate-exclusion cages around a subset of the patches in the adjacent area.

After we had tracked the fate of those seedlings for one year, our data revealed a fascinating trade-off. We found that *lama* survival was highest in the caged adjacent area patches (61 percent), intermediate in the preserve (34 percent), and lowest in the unprotected adjacent area (10 percent). Many of the seedlings growing in the effectively fountain grass–free preserve had grown considerably (for a *lama*, that is; the average height of the two-year-old nursery-grown *lama* "trees" we had transplanted in that first outplanting project at Ka'upulehu was less than four inches!) and had put out one or more sets of true leaves. However, most of the *lama* seedlings in the fountain grass–dominated adjacent area had barely grown, and 85 percent of them still had no true leaves at all.

In other words, from a *lama* seedling's perspective, as long as you were protected from ungulates, being in all that grass was both good and bad: your chances of staying alive there were nearly twice as great as in grass-free soil, but the catch was you probably wouldn't grow much. We later performed additional physiological research on these seedlings that suggested a potential underlying causal mechanism behind this trade-off. These data showed that the extremely low levels of available light down in all that grass appeared to cause the *lama* seedlings to effectively shut down their machinery so that they neither grew (no carbon assimilation via photosynthesis) nor lost much precious water (no transpiration via stomatal conductance). This led us to wonder whether we might be able to facilitate the regeneration of *lama* and other native dry forest canopy trees by creating an intermediate niche with just the right balance of light and soil moisture.

Another insight of potential academic interest and practical value that we gleaned while collecting those additional census data was the seemingly critical role of the native vines. We had observed how these vines came down from the trees and marched across the preserve during the previous wet spring, but we had never considered using these vines in a scientific experiment or as a restoration tool until we observed their performance during the following dry year. Whereas virtually all the other plants appeared to be hunkered down and merely trying to survive that extended drought, two native vine species continued to spread across the exclosure and *increased* their coverage by another order of magnitude. When I noticed one day that a group of native shrubs were doing just fine even though they were completely blanketed by *Canavalia* vines, I suddenly thought, "Hey, maybe that's how things used to work in these forests! And those vines and shrubs are tough if you give them a chance—maybe they should be the foot soldiers to hit the beachhead after we've cleared out the fountain grass."

Shortly after completing that final census, we launched a new "seeding experiment" to test and further explore some of the hypotheses generated by these and other observations and experiments. In contrast to all our previous work at Ka'upulehu, this time we had the luxury to "do it right" because this was a purely scientific experiment funded by our National Science Foundation grant (we even had several specific a priori research questions and hypotheses).

After many hours of what Steve Weller only half-jokingly referred to as our "migraine-inducing discussions," we decided to focus this experiment on the interactions of what we believed were the four most important variables we could feasibly manipulate in the field: water, seed dispersal, vegetation, and available light.

We had repeatedly and often painfully proven to ourselves that, like it or not, at least in this highly degraded and drought-prone ecosystem, we had to supply supplemental water to facilitate seed germination and subsequent plant establishment. But since it was difficult and expensive to establish irrigation systems and deliver the water, we wanted to use this experiment to learn more about how best to use this precious resource.

Similarly, because outplanting individual potted specimens that had been propagated in an off-site nursery was such a slow, laborious, and expensive technique, we wanted to know whether we could more effectively restore dry forests by facilitating on-site, direct regeneration from seeds. We knew that, with a few important exceptions (primarily the vines and a few "weedy" species of native shrubs), the regeneration of the native dry forest flora in general, and the canopy trees in particular, depended on the direct input of fresh seeds. That is, there did not seem to be a viable native soil seed bank—fairly quickly after ripening, these seeds apparently either germinated, died, or were eaten.

Given that the entire native avifauna that once must have dispersed the large, fleshy fruits of canopy trees such as *lama* were now extinct, we were not surprised to find that virtually all of the fleshy fruit seedlings we observed during that wet spring had germinated in the shade of their presumed parents. Indeed, every single one of the thousands of *lama* seedlings we saw then was growing directly beneath a branch of a mature female *lama* tree. We also occasionally saw piles of rotting fruit near the trunks of a few of the larger *Pouteria sandwicensis* trees in the preserve. The Hawaiians once used the sticky sap produced by this member of the Sapotaceae family to catch small birds, and presumably some large bird or birds used to eat the tree's large, sticky yellow to purplish-black berries and scarify and disperse the seeds. Even though the seeds we extracted from these fruits usually looked okay, we never saw a single *Pouteria* seedling at Ka'upulehu, and thus I wondered whether the days of this still relatively common yet senescent tree were numbered.

I really wanted to include this species in our research and restoration projects—there was something about its stately reddish-brown leaves and quiet dignity that immediately made it one of my favorites—but we never did because its seeds were both scarce and difficult to germinate. I also fantasized about importing some large dodo-like birds to see whether they might effectively disperse those rotting *Pouteria* fruits and scarify their seeds. (One member of the working group did try to feed some to his chickens, but apparently they weren't interested.) Given these dead ends, we decided to see whether and to what extent this general lack of seed dispersal and scarification was limiting the regeneration of some key native dry forest species. We proceeded by collecting, processing, and scattering their seeds in a manner we hoped would be functionally equivalent to what the birds used to do.

The third variable we investigated was how the presence of other plants might affect the establishment of the species we seeded into this experiment. This idea was partially inspired by watching all those *lama* seedlings persist inside our caged plots within the thick fountain grass outside the exclosure. We were further motivated by our observations that several other native species similarly appeared able to germinate and grow within patches of existing vegetation at least as well as, if not better than, they did in the more open, vegetation-free areas. We also wanted to know whether it mattered if this existing vegetation was created by native or alien plants.

Finally, we wanted to learn more about how the amount of available light affected the germination and establishment success of both the native and alien plants. We knew from multiple lines of evidence that the native plants in general, and the canopy trees in particular, seemed to prefer the relatively cooler and moister subcanopy niches. For example, the data from that first outplanting project at Ka'upulehu showed that the survival of the plants in the two shaded plots was still significantly greater than the survival of the plants in the two full-sun plots. Moreover, greater than 95 percent of the shaded endangered canopy trees in this experiment were still alive two years after outplanting, and these species as a whole had grown substantially larger and appeared healthier than their full-sun counterparts. Yet we knew little about the causal mechanisms behind these sun-versus-shade differences, and we didn't know whether these patterns would hold for the species that had failed to regenerate on their own during that wet spring.

We hoped to start unraveling and answering these interlocking questions by comparing the performance of plants in the full-sun and shaded niches under different combinations of our other three experimental variables. For instance, if we sowed and watered *lama* seeds in shaded areas that contained other herbaceous vegetation, would they establish as well as, or perhaps even better than, nonirrigated *lama* seeds in shaded areas with no ground-level vegetation?

To combine all four of these variables into a single experiment and still be able to statistically analyze and tease apart their individual and interactive effects, we calculated we would need sixty-four one-square-meter plots. We distributed these plots throughout the preserve in what is called a randomized factorial block design of three treatments: light (canopy shade versus full sun), water (supplemental irrigation versus ambient precipitation), and weeding (all emerging alien species removed versus weeds not removed). We "blocked" these plots into sixteen groups of four; then we located eight blocks in the shade of eight different canopy trees and the remaining eight blocks in nearby open, full-sun areas. Within each block, we randomly assigned two plots to receive the supplemental water and two plots to receive the weeding treatment, in a fully crossed (i.e., factorial) design. Thus, each block contained a plot that would receive supplemental water and be weeded, a plot that would receive supplemental water and not be weeded, a plot that would receive only ambient water and be weeded, and a plot that would receive only ambient water and not be weeded.

Once we had established all these blocks and plots, and clearly labeled the specific combination of treatments for each plot (both to keep ourselves from getting confused and to keep well-intentioned volunteers from weeding and watering our control treatments), we sowed seeds from six different native species into all sixty-four plots and marked their locations within each plot so that we could distinguish their seedlings from others that might germinate from seeds we did not sow.

In addition to finally being able to do some "pure" science, we were able to use the portion of our NSF grant that had been allocated to pay my salary and travel expenses (I was still a postdoc on Kaua'i when we submitted that proposal), along with some matching funds from my new employer, the USDA Forest Service, to hire two scientists. Susan Cordell, a postdoctoral research fellow, was a plant physiological ecologist who had just completed her PhD at the University of Hawai'i and now lived in Hilo; Lisa Hadway was a research technician who had recently completed her master's degree at the University of Hawai'i on the ecology of native Hawaiian snails and now fortuitously lived in Kona. I also felt very fortunate to work with Don Goo and Alan Urakami, two part-native Hawaiian Forest Service technicians with deep roots on the Big Island and extensive knowledge of and practical experience with native and alien plants.

Working on that seeding experiment and several other research and restoration projects with this talented and dedicated team turned out to be among the most productive and fun fieldwork I have ever performed. Because the long arm of the federal bureaucracy hadn't quite caught up with us in those days, we were free to base our work schedules on the demands of the research rather than on a series of perhaps well-intentioned but often counterproductive government regulations. Thus, we could load up the Forest Service van and drive over to Kona from Hilo, check into a cheap hotel, and stay for several marathon days of fieldwork. Sometimes we would rise in the early morning darkness, drive up to the preserve, and collect woody plant tissue samples by flashlight for Susan's predawn native flora water potential study—extracting and quantifying the amount of water within this woody tissue before the sun rose and photosynthesis and transpiration began provided a reliable measure of the amount of water available to plants. Being in the preserve under those cool, tranquil, pitch-dark conditions revealed a whole new level of depth and complexity to that tiny forest I thought I knew so well.

In addition to this core group of coworkers, I had the privilege to interact with and be informed by a growing circle of colleagues and coconspirators; whenever my dream of restoring dry forests seemed hopeless or misguided, it was usually these collaborations and friendships that inspired me to keep going. Through my association with institutions and organizations such as the Forest Service, the University of Hawai'i, several state agencies and nonprofit groups, and of course the North Kona Dryland Forest Working Group, I also had many opportunities to show the Ka'u-pulehu Preserve to, and work with, individuals and groups ranging from young children to native Hawaiian teenagers to distinguished senior research scientists. Some of these interactions turned out to be among the most moving and enriching experiences of my life.

Soon after launching our seeding experiment, I realized that it would have to be my last science experiment in that six-acre exclosure—there were just too many people trying to do too many things in too small a space. I also felt that because we had accumulated a sufficient amount of both academic and more practical knowledge from our smaller-scale experiments and observations, we were now ready to move on to bigger and more ambitious projects.

Around this time, an increasing number of people within and beyond the Hawaiian scientific and conservation communities were beginning to argue that it was time for us to move beyond our conventional small-scale experiments and projects. These people proposed that we "get serious" by scaling up our efforts and conducting research and restoration programs composed of coarse, "blunt tool" treatments that could be feasibly replicated across hundreds or even thousands of acres. I was sympathetic to these pleas, and only too eager to stop merely thinking and talking and writing about the urgent need to scale up our work at Ka'upulehu and start doing it before it was too late.

Fortunately, Potomac Investment Associates (PIA), a real estate development company that leased a large chunk of land in the Ka'upulehu region of the Big Island, had a few employees who were active members of the working group and were also personally interested in dry forest restoration. With their support, the group decided shortly after we launched the seeding experiment that its next major project would be to fence, survey, and then attempt to restore a seventy-acre parcel of PIA land directly across the highway from the six-acre Ka'upulehu Preserve.

This obviously would not allow us to work at the scale of hundreds to thousands of acres that many of us dreamed about, but given the generally indifferent, if not hostile, attitude of the ranching community that dominates the leeward side of this island, we felt fortunate to have an opportunity to work at this intermediate scale. (Contrary to what even many residents of Hawai'i think, the USDA Forest Service does not own or manage any land within this entire archipelago, so working in a national forest was not an option because such places do not exist.)

As with most other projects in this landscape of lava and fountain grass, the first step we took in the new seventy-acre parcel was to use bulldozers to create a few primitive interior access roads and prepare the site for the all-important perimeter ungulate fence. While this work was in progress, a few intrepid colleagues and I spent several brutal days traversing the steep, rough area to establish and census a series of permanent monitoring plots. Sadly, we found that after we descended through a narrow belt of relatively high quality remnant native forest just below the highway, the parcel steadily degraded into the familiar landscape of scattered islands of senescent trees within a sea of fountain grass.

The highlight of our otherwise bleak botanical survey occurred on the second day while we were traversing a savanna-like area that contained only fountain grass and a few diminutive and senescent trees. I was trudging along with my head down, nursing my heavily scratched and bruised body and straining to carry my pack full of rebar stakes, when one of my sharp-eyed colleagues suddenly stopped and pointed to a gnarled old *lama* tree below us. "Is that what I think it is?" she whispered.

We dropped our stuff and waded through the grass until we reached that tree. I had no idea what was so special about this particular *lama* until I saw a liana twined around a section of its canopy. It had large, oblong, leathery leaves, and little white morning glory–like solitary flowers. I traced its spreading branches down to a few tough, cable-like stems that looked as if they had been there since the island first erupted out of the sea.

"Yup, *Bonamia menziesii*—good spot!" another member of our team proclaimed after an up-close inspection. "I don't think there are any known specimens of this guy anywhere on this side of the entire island. Bob, add this to your list of federally endangereds at Ka'upulehu. Looks like there may even be some ripe fruit up there!" He scampered up into the canopy and collected a handful of small capsules to take to another colleague who had an almost mystical knack for propagating rare Hawaiian plants.

I stepped back and looked at what appeared to be a healthy, even vigorous plant clinging to this tree as if to a lifeboat. What had this forest been like when it first germinated and began searching for something to climb? Had there once been an abundant and ecologically important *Bonamia* population at Ka'upulehu, and perhaps in the other dry forests across the archipelago? Was it merely blind luck that had enabled this particular specimen to survive all this degradation? I knew that unless plants learned to talk, or we became infinitely better at understanding their language, these kinds of questions would almost certainly remain unsolved mysteries.

When our survey was done, we found that aside from a few vines and shrubs within the relatively intact forest just below the highway (why this strip of vegetation had survived was yet another unsolved riddle), the entire seventy-acre parcel contained virtually no native understory plants. Yet, much to my amazement, almost immediately after the bulldozers completed their work, some native shrubs, herbs, and vines began popping up along portions of the newly created crushed-lava roads. Over time, some of these sections contained relatively dense and diverse plant communities, and although most were composed of both native and alien species, they tended to be dominated by the native vines and shrubs.

I assumed that some of these native species must have germinated from seeds that had recently matured in the six-acre preserve across the highway and then had been dispersed in by wind or perhaps birds. This seemed especially plausible for plants such as 'ilima (*Sida fallax*), an extremely variable member of the Mallow family whose yellow flowers are often used for leis, and the beautiful white Hawaiian poppy (*Argemone glauca*)—one of the very few endemic species that still has thorns, perhaps because of its fairly recent establishment within the archipelago—because both of these relatively common species produce abundant crops of small, lightweight seeds. But how did the big, heavy seeds of the *Canavalia* vines get here? Had the dozers unearthed and scarified a cache of ancient *Canavalia* seeds? Down under all that fountain grass, was there a viable soil seed bank for some native species after all? Were these new, barren corridors of lava created by bulldozing ecologically analogous to the "roads" of molten lava that once flowed through these dry forests?

As usual, nobody knew the answers to these and other important, relevant questions. Yet, despite how fascinating many of my colleagues and I found such questions, and how much we enjoyed formally investigating and informally speculating about them (ideally after a few beers), I was beginning to realize that our ignorance on these matters need not prevent us from attempting larger-scale restoration projects in these and other highly degraded ecosystems.

In fact, most of the more formidable barriers that had so far prevented us and many other colleagues from scaling up our efforts had more to do with logistics, economics, and politics than with unsolved ecological and evolutionary riddles and limitations. For instance, we already knew that it would be futile to try to restore this landscape without first controlling all that smothering and fire-promoting fountain grass. But in order to accomplish this, we needed some kind of coarse, blunt tool that could handle fountain grass more efficiently than our effective but slow, expensive, and often treacherous weed-whack, backpack-spray routine.

The only potentially plausible larger-scale fountain grass control techniques we had thus far come up with were aerial applications of herbicide, intensive ungulate grazing, controlled burns, a biocontrol agent or agents, and US Army tanks. Unfortunately, none of these ideas had panned out.

Not long after we finished fencing the new seventy-acre parcel, we finally got the necessary funds and permissions to spray a grass-specific herbicide over the entire area by helicopter. Sadly, however, my subsequent censuses showed that the aerial application had largely failed to kill or even suppress the fountain grass within the exclosure; consequently, we never employed this technique again.

Despite lots of optimistic talk among some members of the working group about the potential for using cattle and goats to graze and trample fountain grass via a series of fenced paddocks, we never tried this technique because the logistics were always too complicated and the economic costs and ecological risks too great (it would have been a total disaster if those animals ever got inside our exclosures).

Using fire to control fountain grass similarly wound up being too costly and risky to attempt, even on an experimental basis. The Forest Service fire experts I brought in from the mainland informed us that given this grass's enormous bulk and extreme flammability—I found on average over twenty tons of dry aboveground fountain grass biomass per acre within the seventy-acre exclosure—it would require far too much time and money to safely perform a controlled burn. There was also strong opposition to this approach among some members of the working group who had previously witnessed disastrous "controlled burns" on the Big Island.

Our vision of using a biocontrol agent also turned out to be a mirage. Although the Forest Service operated a biocontrol facility on the Big Island, my colleagues there claimed that even if it was their top priority, it could take well over a decade of intensive work and cost more than a million dollars to screen and develop a potential biocontrol agent or agents, which might or might not put a meaningful dent in all that fountain grass. Furthermore, they explained that because of the economic importance of grasses in general (many of our most important agricultural crops are grasses), to the best of their knowledge no one had ever attempted to use biocontrol agents to control a grass species before. (There was in fact another grass on the Big Island in the same genus as fountain grass, *Pennisetum clandestinum*, or *kikuyu* grass, that the ranchers considered an important forage crop for their cattle.)

Finally, our dream of using army tanks to crush and suppress fountain grass turned out to be just that. There was an army base on the island that ran tanks across some pretty rough and rugged terrain, and we discussed the idea of using some of them with two army biologists who worked at that base and were also members of our working group. Not surprisingly, however, the logistic, political, and economic realities associated with getting those tanks to Ka'upulehu and using them for ecological restoration once again proved insurmountable.

A few weeks after our failed aerial herbicide experiment, I sat down for lunch on a big lava rock next to our new interior access road in the seventy-acre exclosure and looked out despondently at the vast expanse of treeless fountain grass that lay between me and the ocean. But then I looked down at an impressive patch of native shrubs and *Canavalia* vines growing on that access road right under my nose. Despite the heat and prolonged drought, the vines sprawled down that road like a thick green snake for at least another seventy-five feet. A Kona farmer had recently told me that his coffee plants took up far more water from the crushed lava substrate left behind by bulldozers than they did from land cleared by other means. Could this also explain why those vines and other plants were doing so well out here?

I studied the huge piles of torn-up fountain grass clumps and lantana shrubs that the dozer operator had created to make this road and suddenly thought, "Who needs army tanks when we've already got bulldozers right here!" Could they be the blunt tool we've been seeking for so long? If the farmers used them to clear land for their crops, the ranchers used them to put in their fences, and the developers used them to create their subdivisions, why couldn't we use them to scrape away the fountain grass and prepare our sites for dry forest restoration projects?

When I excitedly presented this plan at the next working group meeting, responses from the other members ranged from "Great idea!" to "Over my dead body! Bulldozers are evil—look at the ecological destruction they've caused in Hawai'i and elsewhere! Are you saying that we have to destroy the dry forest in order to save it? Even if it 'worked' in the short term, you have no idea what the long-term consequences of dozing might be!"

I responded by arguing that (1) like all tools, bulldozers could be used for both good and evil purposes; (2) I wanted to doze only already "destroyed" areas with no native species present; (3) we and others were already using bulldozers for roads, fences, and firebreaks; (4) we didn't know the long-term consequences of any of our other management actions; and (5) none of our other potential large-scale techniques had worked, no one else seemed to have any other ideas, and the clock was ticking! This in turn led us back to some old, more general topics such as the pros and cons of doing nothing versus trying risky and potentially disastrous interventions, and what our group's overall mission, guiding philosophy, and general course of action were or should be. After many subsequent debates stretching across several meetings that included the often conflicting opinions of outside experts, I finally received permission to cautiously try bulldozing on a small, experimental scale.

I wanted this new experiment to be as "big picture" as possible—I had had enough of doing such things as counting the number and measuring the area of leaves on hundreds of individual plants in dozens of tiny plots. Instead, I planned to step back and collect the kind of bottom-line data that might be more valuable to people who wanted to attempt their own restoration projects. I therefore wanted to investigate how the native and alien plant communities as a whole would respond to different treatments that could conceivably be adapted and applied to real-world, larger-scale restoration projects. For example, by the end of the experiment, I wanted to be able to tell an interested group or individual landowner something like the following: "Our data suggest that an optimal restoration strategy would be to start by bulldozing off the fountain grass and then seeding in native understory species X, Y, and Z. We found that dozing was 65 percent cheaper and 85 percent faster than weed whacking and spraying herbicide on the grass by hand, and all but one of the native species we seeded into our experiment were significantly larger and more abundant in the dozed areas than they were in any of our other fountain grass control plots. About six months later, depending on the weather, you should be able to reduce your irrigation to X gallons per week and start outplanting canopy tree species A, B, and C in the shade provided by your established understory shrubs and vines. The best way to control fountain grass and the other alien weeds that will start popping up around this time will be to . . ."

In addition, I believed we already knew enough to conclude that at least on this island, if not in the entire Hawaiian archipelago, any effective dry forest restoration program must include ungulate exclusion, control of the dominant alien species, supplemental irrigation, and seeding or outplanting, of key native species. Thus, unlike our other research projects, this experiment would not employ a full factorial design: there would be no "unfenced area," "intact fountain grass," "ambient water," or "unseeded" replicates.

Yet the more I thought about it, the more I realized that even if we had a fleet of bulldozers and several thousand acres of degraded land at our disposal, it still wouldn't be possible to perform meaningful research or applied restoration at this coveted landscape scale. Even if we had the support and cooperation of the local landowners and the resources to do whatever we wanted, what exactly *could* we do across such a large area?

Even if we could kill or suppress fountain grass on a landscape scale, we knew that native dry forests would not magically rise up out of all that newly exposed soil and lava rock on their own. Instead, we would almost certainly see the eventual reestablishment of fountain grass and various outbreaks of other formerly suppressed noxious weeds. Unlike the high-quality six-acre Ka'upulehu Preserve, this entire region of the island had virtually no significant populations of regenerating native understory plants. In addition, we knew that whatever we used to kill the fountain grass would almost certainly kill the few natives out there.

Thus, it was pointless and perhaps even counterproductive to attempt to control fountain grass without simultaneously implementing additional restoration treatments to facilitate the subsequent establishment of native plant populations. Consequently, after extensive research and brainstorming, we decided to build this experiment around a single multifaceted question: How do light availability (full sun and 50 percent shade); grass control treatments (bulldozing, weed whacking and herbicide application, use of plastic mulch, and trimming); and native species additions (outplanting and direct seeding) affect the establishment of native plants and the suppression of fountain grass?

As I thought through the nuts and bolts of what I now called the Big Experiment, I realized that its spatial scale would be constrained by four major factors. First, of course, we didn't have unlimited time or money or the autonomy to do whatever we wanted—we had just a few years to design, implement, and terminate an experiment that had to be acceptable to the other members of the working group. We also had only a modest amount of money remaining in our research budget and the part-time labor of just a few people—Susan, Lisa, Don, Alan, and me—and all of us were struggling to keep up with our other projects and responsibilities.

Second, given the political realities of working on the Big Island, the only feasible arena in which to conduct this experiment was our seventy-acre exclosure. No agency or individual was going to let us drive off and fence out the ungulates from their land and then try to start growing a forest that might take hundreds of years to mature. Indeed, despite years of lobbying from the environmental community and some of its own administrators and biologists, local and state politics had recently led the state's Division of Forestry and Wildlife to reject a bid by The Nature Conservancy and others to preserve and restore North Kona's vast yet horribly degraded Pu'u Wa'awa'a Ranch—parts of which the eminent botanist Joseph Rock had described a century earlier as "the richest floral section of any in the whole Territory." Instead, the state had decided to continue leasing much of its Pu'u Wa'awa'a "forest preserve" to a local rancher.

The third major factor constraining the scale of the Big Experiment was the physical and logistic difficulty of working in this parched and rugged landscape. For instance, since we knew we needed, at least initially, to provide our plants with supplemental water, we obviously could not create an experimental arena larger than we were capable of irrigating. Yet our only source of water was an irrigation tank just below the highway at the top of the seventy-acre exclosure, which we had recently installed for general restoration and firefighting purposes. We knew from past experience that the amount of water and pressure we could expect this tank to deliver via our gravity-driven network of pipes was going to be modest at best. We also knew that purchasing, transporting, and installing all that plumbing and various other research equipment and supplies would create many formidable, time-consuming, and expensive logistic challenges.

The amount of accessible, homogenous space within the seventy-acre exclosure was yet another important physically limiting factor. After much scouting, I decided that the only feasible arena for this experiment was a relatively accessible flat, treeless area at the bottom of the exclosure that contained only fountain grass and lantana shrubs. Yet, as I had discovered in the six-acre preserve, hidden beneath this seemingly uniform veneer of fountain grass and lantana was a far more complex mosaic of different soils, slopes, and lava substrates.

Both because of this ecological heterogeneity and for various logistic reasons, we decided to arrange our experimental replicates into discrete pairs of parallel and adjacent strips of shaded and unshaded plots. In theory, this "randomized paired block design" allows researchers to control for the overall variability of their field sites by dividing them into smaller, relatively consistent subsections. Thus, we could place one set of shaded plots and one set of unshaded plots in an area that contained, say, relatively thick and tall stands of fountain grass; pair the second set of shaded and unshaded plots in another area with deeper soils; and so on. Yet in practice we found that the spatial scale of these "uniform" subsections was far smaller than we would have liked, and in some cases the combination of this spatial heterogeneity and the complexities of implementing the fountain grass control treatments (e.g., bulldozing one plot without running over the nondozed adjacent plots) forced us to locate some "paired" plots in physically and ecologically distinct areas.

The fourth and final major limiting factor was availability of native seeds. One of our primary goals was to establish an abundant and diverse native understory population that could coexist with, or perhaps even suppress, the fountain grass and other alien species that would inevitably recolonize each plot, no matter how effective our initial control treatments were. To even have a chance of achieving this goal, we would have to add a substantial number of native seeds and outplants.

Until we started planning the Big Experiment, I did not fully appreciate the fact that our little six-acre preserve was about the only place left to
collect local seeds from most of the native dry forest species. Although
there were still a few patches of remnant forests scattered across this side of
the island, even if we had the landowners' permission to access them
whenever we wanted (a big if), we simply didn't have the time or person-
nel to repeatedly slog out and try to collect a handful or two of ripe fruit before the rats and birds got them. And because none of these patches had
been fenced to exclude ungulates, from what I could tell from the highway and a few discreet little reconnaissance missions, the diversity and
abundance of their native understory floras generally ranged from sparse to
nonexistent.

In addition, unlike the situation in temperate areas, seed production in
the tropics tends to be episodic and weather dependent. We also found that
the viability of seeds produced by some of the key dry forest species was extremely variable and unpredictable. For example, almost all of the *lama*
seeds that Dave and Tim collected during their original floristic survey of
the Ka'upulehu Preserve germinated and grew into healthy seedlings.
However, almost none of the *lama* seeds from some (but not all) of our
later collections germinated, even though all the relevant ecological and
methodological variables were, as far as we could tell, identical to those of
the first collection. Consequently, not only was it difficult to estimate how
many seeds from each of our target species would be available for the Big
Experiment, but it was also virtually impossible to know how many of the
seeds we did collect would be any good. Thus, even if we weren't limited by
all the other constraining factors, we still could not work at a landscape level
because we would never be able to collect enough seeds and grow enough
plants to populate more than an acre or two of land.

Monday, May 17, 1999, finally arrived. After six months of intensive planning and preparation, it was launch day for the Big Experiment. I shoved
the last cooler into the backseat of our spiffy new Forest Service Ford Ex-
pedition and took one final look at my list to convince myself that everything we needed was in the van. Satisfied, I climbed in and pulled out of
the base yard.

The sun was just peeking above the horizon and breaking through the
low-hanging clouds over the ocean, and the ethereal dawn light made the
volcanoes and Hilo Bay look like one of those garish, overpriced rainbow-
mermaid-dolphin paintings that the tourists snap up in Waikīkī.

Two hours later, as I approached the west side of the island, I spotted
a *pueo* (*Asio flammeus sandwichensis*), the Hawaiian short-eared owl,

hovering over the grassland alongside the highway. I always felt inspired whenever I saw a *pueo*—it was uplifting to see a native bird that was actually doing well—and I hoped that its presence that morning would prove to be an auspicious omen.

As I headed down the Mamalahoa Highway toward Kaʻupulehu, I saw that for once the air over North Kona looked relatively clear and free of vog (volcanic smog—an irritating mixture of water vapor, carbon dioxide, and sulfur dioxide). I turned into the seventy-acre exclosure, bounced and skidded down the steep and eroding perimeter access road, and, at the bottom, parked behind Lisa's pickup truck. As usual, she had done a superlative job of organizing everything, so there was little for us to do except wait for everyone else to arrive.

We walked over to take one last look at all our soon-to-be-transplanted species in their long plastic containers. After Don and Alan set up a makeshift nursery here, we had brought all the plants over from our Hilo greenhouse so they would have a month before the start of this experiment to acclimate to the harsh Kaʻupulehu climate. We had selected a dozen different native species (three canopy trees, including two endangered species; six shrubs; two vines; and one herb) on the basis of their local seed abundance and their ability to germinate and grow in the field. To compare the results of direct seeding versus outplanting, we decided to sow all twelve species directly into each plot and transplant individuals from as many different species as we could propagate in sufficiently large numbers (because of limited seed availability and poor germination and growth, we ended up with healthy seedlings from only seven of these twelve species).

While we looked at the more than 1,400 plants in front of us, all neatly grouped and labeled with their designated outplanting area, I thought about the countless hours we had spent collecting and processing seeds and nursing the subsequent outplants. As with just about every other aspect of our work at Kaʻupulehu, I could convince myself that these benches of plants, and this experiment as a whole, represented the culmination of either a noble, heroic triumph; a tragic, quixotic waste of time and resources; or, perhaps, something in between.

I walked past our plots and up the steep slope that bordered them to snap a few final pictures of our overall experimental arena. The net result of all those interacting constraining factors was that the Big Experiment occupied a section of this exclosure measuring roughly 150 by 300 feet. Stretched out below me lay the four blocks that formed the core of the experiment. Each block consisted of two parallel strips each 100 feet long by 30 feet wide, with about 6 feet of space between the strips within a block and 10 feet of space between adjacent strips of different blocks. We covered one randomly selected strip in each of these four blocks with a layer of 50 percent shade cloth. Then we bungee-corded its grommeted ends to a supporting framework of telescoping metal poles that allowed us to maintain a constant height above the uneven ʻaʻā lava and fountain grass substrate. Each strip in turn contained four plots measuring 20 by 20 feet and spaced 6 feet apart from each other and the edge of the strip, for a grand total of thirty-two plots (four blocks, with two strips per block and four plots per strip).

We randomly distributed our four different fountain grass control treatments among the four plots within each strip so that each plot received a different treatment. We implemented the bulldozing treatment by scraping the blade of a Caterpillar D8 backward across the surface of each of the eight plots assigned to receive this treatment. This action, which took only a few minutes once we got the dozer in position, removed the top ten or so inches of lava substrate, soil, and fountain grass.

Our second fountain grass control treatment was the tried-and-true weed whacking and herbicide routine. I initially didn't want to include this treatment because I felt we had enough experience to know its strengths and limitations, and I didn't think it could be feasibly implemented on a larger scale. However, as our experiment progressed from my initial grandiose dream toward its much more humble final reality, I began to realize that, at least for the foreseeable future in Hawaiʻi, this project *was* large-scale dry forest research and restoration.

Our third fountain grass control treatment was a layer of heavy, 100 percent light-blocking black plastic mesh. I had read and heard about using plastic to smother and "solarize" (cook) weeds and the underlying soil seed bank in other ecosystems and hoped it might be an effective methodology in our hot, dry climate as well as a potentially more palatable technique for those who preferred not to use herbicides. To implement this treatment, a few months earlier we had simply weed-whacked the grass to ground level and then covered the entire plot with plastic.

Our final fountain grass control technique was what we called the "trim treatment." While I felt we had more than enough evidence to justify not including a true control treatment in this experiment (i.e., plots with intact, untreated stands of fountain grass), I had wanted to try some kind of thinning technique ever since we first saw all those *lama* seedlings within the intact stands of fountain grass just above the six-acre exclosure. In addition, one of the most interesting and surprising results of our ongoing seeding experiment was that the diversity and abundance of the native species appeared to be *greater* in the nonweeded plots. Given that some of these nonweeded plots were covered with a dense layer of alien and some unplanted native vegetation, this result once again suggested that at least some of our target restoration species might be able to grow and establish within only partially controlled stands of fountain grass. To test this hypothesis, we decided to weed-whack the grass down to twenty inches one week before we launched the Big Experiment.

I walked back down the slope and joined Lisa under one of the shade cloth strips. In the previous week, we had subdivided each of the plots within each strip into four evenly sized quadrats and randomly selected one quadrat in each plot to receive our native outplants. We then designated the quadrat located diagonally across from this outplant quadrat to receive our batches of native seeds; the remaining two quadrats in each plot would serve as controls for these treatments (no native species added).

Even though I knew some biostatisticians wouldn't like this "restricted randomization" design, I thought setting things up so that the quadrats to which we added seeds and outplants would always be adjacent to and bordered by the quadrats without added native species would be an efficient way to track the spread and coverage of both the native and alien plants over time. I also thought this design might yield some practically relevant and valuable information. For example, we would (I hoped) be able to answer questions such as "Did the *Canavalia* vines originating from the outplanted quadrats spread into and cover the adjacent quadrats without added native species more than did the vines originating from the direct-seeded quadrats?" Or, conversely, "Did the outplanted or seeded quadrats more effectively prevent or retard the invasion of alien weeds from the other quadrats in their plot?"

I looked at the army of color-coded pin flags that ran down this cloth-shaded strip—marking off each plot, quadrat within plot, and outplant, direct-seeding, or control area within each quadrat—and wondered how much of this spatial design might be intelligible to the volunteers who, we hoped, would be arriving shortly. What would they think of the rationale and motivation for this experiment as a whole? Would they look at all this and be inspired, or would they think we were nuts?

Our plan was to use these volunteers primarily to transplant the forty or so plants designated for each outplanting quadrat. The so-called Kona planting method that we wanted them to employ consisted of creating or exploiting cracks in the lava with metal digging bars and then mixing just enough premoistened commercial potting soil into these cracks to keep the seedlings upright. Because the substrate of several of the outplanting quadrats was essentially an unbroken sheet of 'a'ā lava, it would require considerable strength, endurance, and dedication to penetrate that lava and plant all those seedlings well enough for them to have a fighting chance to survive and successfully establish.

I had participated in enough conservation-oriented volunteer projects to know that they were nearly impossible to predict; sometimes, no matter how much effort was poured into advertising and organizing them, they turned into poorly attended, discombobulated disasters. Yet at other times, a group of highly skilled and enthusiastic people would appear out of nowhere and knock out a project so quickly and thoroughly that I would end up feeling that no environmental problem was too large or intractable to solve.

We walked down the strip and went over the planting details one last time. If there was one person on this island even more eager than I was to have this experiment under way, it was Lisa. On top of her countless other responsibilities and chores, she had spent untold hours here by herself, doing everything from gluing pipes to chasing goats to dealing with some overly aggressive members of Hawai'i's Operation Green Harvest (a statewide, helicopter-driven effort to eradicate marijuana cultivation), who were more than a little skeptical when she explained that all that irrigation equipment and all those bales of potting soil were for "legitimate scientific research purposes."

As more people arrived with their digging implements, work gloves, and good spirits, my anxiety began to evaporate, and I felt increasingly confident that this was going to be one of those good days. By the time Hannah Springer was ready to give her brief overview of Ka'upulehu and bless our experiment, about fifty people had somehow found the time and desire to make it down to this remote and austere spot. I looked around and saw many familiar faces — curious friends, colleagues from the state and federal agencies, members of environmental and educational groups, members of the working group — and several individuals and whole families I had never seen before. Hannah was in her usual fine form and eloquence, and I noticed more than a few moist eyes around the circle when she spoke about what North Kona's once mighty dry forests meant to her ancestors, herself, and her children.

Then it was all over in a flash. Much as with our initial outplanting of those 200 plants in the six-acre exclosure two and a half years earlier, what had taken so much time and effort to prepare took only a few short hours of highly organized chaos to execute. Once again, I was amazed at how hard even the people who were not used to this kind of manual labor were willing to work to get the job done. All of a sudden the benches were empty, the plants were in the ground, and the volunteers were heading back up to their vehicles before we even had a chance to say thanks or offer them one of the excellent beers generously donated by the Kona Brewing Company.

After everyone else had left, my colleagues and I walked around to check the accuracy and quality of the outplantings, but we found almost nothing that needed fixing. Even within the most difficult quadrats, the volunteers had dutifully transplanted the outplants within their prescribed areas, and virtually every single seedling was carefully tucked into and propped up within a decent-looking hole or crack.

Finally, I took out our precious bags of native dry forest seeds and, feeling like some magic fairy, delicately sprinkled their contents across the designated section of each of our thirty-two direct-seeded quadrats. Compared with the outplantings, sowing these seeds was a dream; I was able to seed all the quadrats by myself in about fifteen minutes.

To prepare for this direct seeding, I had been in the lab late into the previous night draining and dividing the seeds that we had found germinated best after a prolonged soak in hot or cold water. Just before I was about to pour one-thirty-second of each plant's total seeds into each bag, it dawned on me that such a collection had almost certainly never been assembled before, and very likely would never be assembled again, before one or more of these species became extinct. I thought of a conversation I had had with a prominent, battle-scarred member of the Hawaiian conservation community shortly after I started working at the National Tropical Botanical Garden. When I asked him what he thought was the most important thing I could do for Hawai'i's native flora, he shook his head sadly and said, "Take good pictures."

Feeling a bit as I imagined a medieval monk must have felt as he sat in a gloomy monastery, carefully copying the contents of manuscripts that few of his contemporaries cared about or even knew existed, I pulled out my camera and snapped what I hoped were unnecessary photographs of all those beautiful seeds.

Shall We Dance? The Trade-Offs of Science-Practice Collaborations and Community-Driven Restoration

A few months after launching the Big Experiment, I spent a day working with the other members of the North Kona Dryland Forest Working Group and some volunteers in the narrow band of remnant native forest that ran across the top of the seventy-acre exclosure. We cut and sprayed fountain grass and then outplanted native species along both sides of a trail Don Goo and Alan Urakami had recently constructed with a group of local kids to provide a relatively safe and easy way for our growing number of visitors to experience this otherwise inaccessible area. As always, it had been a lot of hot, hard work, but it was deeply satisfying for me to get a rare chance to just *do* ecological restoration without having to collect data or formulate and test research questions and hypotheses. After all the pain and suffering fountain grass had caused me, I also derived great pleasure from simply killing it and ridding the land of its pestilence.

Physically working together had proved to be an increasingly important bonding experience for those of us in the working group. These "workdays" allowed us to interact without having to literally and figuratively sit in our customary chairs around the table and speak from the perspectives of our entrenched roles within the group ("the landowner," "the US Fish and Wildlife Service representative," "the local," etc.).

I also relished the opportunity to contribute directly to our applied restoration program and to help build and maintain good relationships between "the scientists" and "the practitioners." It seemed that within and beyond the Hawaiian Islands, such relationships were deteriorating and the science-practice gap was widening. Perhaps this was partly because we all were leading increasingly hectic lives and struggling to get more done with fewer resources in less time. Consequently, some scientists were becoming more conservative with their time and expertise and pulling away from applied collaborations and outreach activities. Some practitioners in turn were resisting the constraints and "authority" of formal science and choosing to do things "their way" instead. Thus, rather than working together, these groups were ignoring or competing with each other; some programs had even developed an either-or attitude: "We are going to do either 'real science' or 'real restoration,' but not both!"

I was also dealing with rising tension in my personal and professional life in general and at the Kaʻupulehu Dry Forest Preserve in particular. Many of the working group members' initially sincere but perhaps naïve support and enthusiasm for science had faded as they came to see its often severe limitations. Unlike the romantic, knight-in-shining-armor way that scientists tend to be portrayed to the public, scientists in the real world rarely have silver bullets to solve complex applied problems. On the contrary, our research often *inhibits* on-the-ground projects without providing any immediate practical benefits. In addition, I was starting to feel as if one of my major roles as a scientist at Kaʻupulehu was to throw a wet blanket on what I perceived as the reckless desires of some of the group's members to implement ever more projects with ever less care and planning.

Yet at the same time, unlike many of my scientific colleagues, I was deeply caught up in the world of applied conservation. Consequently, I had more empathy for the "just do it" urgency of the practitioner mentality. Trying to continually balance these two opposing forces within myself and help bridge the science-practice gap within organizations such as the working group was wearing me down and burning me out. When I noticed myself starting to act and feel like some of the more cynical, cheerless, defeated members of the conservation community I swore I'd never become, I knew it was time for a break.

Two months later, I came back from a three-week trip to the mainland feeling refreshed and energized. I had deeply enjoyed attending some meetings, giving a few research seminars, talking with old colleagues and meeting new ones, and squeezing in some much-needed personal vacation time. But while it had been great to get away, it was even better to come back to Hawaiʻi with a renewed appreciation of what a paradise these islands are, and why trying to preserve and restore their remaining native species and ecosystems was worth all the struggle and strife.

After a week of catch-up in my office in Hilo, I was eager to get over to the other side of the island and see how things had progressed at Kaʻupulehu in general and in our Big Experiment in particular. It had been five months since we launched that experiment and nearly eight weeks since I had seen it in person.

Since the road leading to the bottom of the seventy-acre exclosure had become nearly impassable, I decided to park the van just below the highway and walk down to the Big Experiment via Don and Alan's trail. I entered the forest and stopped for a moment just to breathe in the cool, fresh air and admire the strange but beautiful trees. The more I got to know this ecosystem, the more I understood why so many people throughout the tropical world once chose to settle here.

I was delighted to see how much more fountain grass had been cut and sprayed and native species outplanted in my absence, and it was immensely inspiring to see all those healthy new plants reaching up toward the sun. However, I couldn't help noticing as I walked down the trail that the quality of the outplants was steadily decreasing. In the first, closely supervised upper areas that we had planted with the volunteers before I left for the mainland, every plant was outfitted with its own irrigation line and identification tag. But in the lower, newer, unsupervised planting areas, I saw more and more unlabeled, unirrigated, unhealthy-looking plants.

In virtually all of our scientific experiments at Ka'upulehu, we had to grow our own plants under carefully standardized conditions to ensure that all replicates in each species and treatment combination were derived from the same seed source and approximately the same age and size. For similar reasons, we always tagged each plant and collected and maintained an extensive database of relevant pre- and post-outplanting information.

As a group, we had agreed that it probably wasn't necessary to follow such rigorous and time-consuming protocols for the plants we used for restoration. However, exactly what our minimum standards and record-keeping procedures should be had become yet another contentious science-practice issue. Should we accept and outplant each and every plant that came our way regardless of its quality or lack of background documentation? Should we tag every plant, record whatever relevant information we had or could measure, and collect at least some post-outplanting data, or just get as many plants in the ground as quickly as possible and move on? We largely failed to resolve these kinds of questions, and even when we did at least temporarily approach consensus on some issue, there was often a large gap between what was said around the meeting table and what happened in the field.

I knew that the guys who had done all the additional work along Don and Alan's trail would say that they were just doing the best they could to keep up with our growing backlog of plants. Indeed, once the word had gone out that we had a secure, fenced, and irrigated place to outplant native dry forest species, people started coming out of nowhere and (sometimes literally) begging us to take plants they could no longer care for themselves. When we combined all the donations from the desperate backyard hobbyists, nursery managers, and underfunded state agencies with our own considerable surpluses, we, too, found ourselves with far more plants than we could handle.

When I reached the end of the trail and started walking down the interior access road, I saw that the technicians had attempted to solve this problem by hacking out some crude new outplanting areas within this degraded, largely treeless section of the exclosure. They had also set up several makeshift benches along the road to store some of our surplus plants. Upon closer inspection, I saw that several of these benches were badly overcrowded, most of their plants were untagged, and many were dead or dying.

As I walked past the last of these benches and outplanting areas, my good spirits were starting to fade. The last thing I wanted to do was get in the middle of what was sure to be another lengthy and unpleasant debate over what was happening with our outplanting program, what adjustments should be made, and how these new plans should be communicated and enforced. So I ordered myself to forget about all this for now and headed down to see the Big Experiment.

Even from a few hundred yards away, I could tell that some of the species we had transplanted and seeded into this experiment had grown beyond my wildest dreams, and several plots were now dominated by one or more *Chenopodium oahuense* plants (an endemic species within this widely distributed genus, commonly known as lamb's quarters or goosefoot). This was especially surprising because this was one of the more sparsely distributed, seemingly unimportant plants up in the six-acre Ka'upulehu Preserve; consequently, we almost hadn't even included it in our native species mix in this experiment. Yet down here, it had somehow grown from the scrawny little seedlings we had outplanted into towering treelike shrubs that were now pushing up against our six- to ten-foot-tall shade cloth roofs. Some of these shrubs were also thickly covered with native vines, and an impressive diversity of other native species had established themselves beneath these shrub-and-vine canopies.

I felt like a kid in a candy store as I prowled around and discovered one encouraging scene after another. Even without quantitative data, it was obvious that the native plants beneath the shade cloth structures were doing substantially better than they were in the adjacent full-sun strips. And even though I was admittedly biased in their favor, I nevertheless felt confident that bulldozing was still the most effective grass control treatment by far. Many of the dozed plots also now contained impressive little plant islands composed of most of the native species we had added. Several of the vines in these plots had taken off and now swarmed over a substantial fraction of their adjacent quadrats that had not been outplanted or seeded. "Maybe some seeds from the *Chenopodium* plants and the other fruiting shrubs will blow into these quadrats and establish beneath those vines," I thought. "Then, in phase two of this experiment, we'll bulldoze all the fountain grass out of the surrounding area, come in with the canopy trees . . ."

After I examined about half of our thirty-two plots, my only real disappointment was the "trim" fountain grass control treatment. In retrospect, I wished we had been more aggressive and initially cut the grass all the way down to the ground instead of only to knee height. It had come back faster than I had expected, and it was already overtopping most of the native species. But then, out of the corner of my eye, I spotted some non–fountain grass green in one of the direct-seeded, trimmed quadrats. When I walked over and parted the thick overlying grass, I saw several healthy seedlings of our two federally endangered canopy trees poking up out of the ground.

I sat down beside those delicate little creatures and looked at them for a while. I thought about their precontact history and the string of miraculous events that must have occurred for them to reach these islands, establish themselves, and ultimately thrive here. I thought about the Hawaiians' relationships with these species and tried to picture this landscape as it might have looked during the prehistoric period when this region was thickly settled. Then I looked down at the ongoing construction of Charles Schwab's private golf course a few miles below me. A fleet of bulldozers was pulverizing the lava fields, and a never-ending chain of dump trucks was hauling in topsoil behind them. Despite the precarious existence of these two endangered trees, I knew that if we had even a tiny fraction of Schwab's money, we could almost certainly prevent the extinction of these and the other dry forest species.

The following week, Susan and I drove to Kona to attend another working group meeting and, we hoped, sneak in some science before returning to Hilo. On the way over, she filled me in on developments that had unfolded in the group while I was on the mainland. We joked about the not-so-funny fact that there might now be enough ego clashes, shifting alliances, hidden personal agendas, cultural and stakeholder divisions, and general overall drama to rename our group "As Ka'upulehu Turns."

Like me, Susan cared deeply about the preservation and restoration of Hawaiian dry forests and had repeatedly gone far beyond the call of duty to help further these goals. Yet she too had grown increasingly weary of all the politics and time-consuming tasks associated with being an active and responsible member of this group, and she yearned for more time to focus on her own expanding research program and myriad other responsibilities.

Despite all of the group's internal strife, however, we both felt that it had somehow remained an important and effective organization. In fact, things were now humming along in all three of our major focal areas of dry forest restoration, research, and outreach. On the restoration front, in an impressively short period of time we had managed to fence the entire seventy-acre parcel; survey its flora; install firebreaks, access roads, and irrigation lines; cut and spray several acres of fountain grass and kill some other key noxious weeds; and outplant thousands of native plants. Our growing number of grants and increasing support from other agencies and institutions had also finally enabled us to actually *pay* a few people to do some of the never-ending on-the-ground work and help organize and run our volunteer programs. Thus, there was good reason to believe that our future restoration accomplishments would, if anything, be even more impressive than what we had already accomplished.

In addition, over time I had noticed that despite all the tensions and conflicts, some valuable synergies had developed among these three not-so-separate focal areas. For example, all this restoration work now formed a major component of our outreach programs. Many of our volunteers wound up absorbing an impressive amount of knowledge, which we hoped would lead at least some of them to educate and inspire others to help save Hawai'i's remnant dry forests. The fruits of our restoration efforts also provided much of the content of our outreach materials and presentations. Rather than simply recite a bunch of grim statistics, we could now focus on our accomplishments and our growing confidence that this ecosystem could be at least partially restored. Finally, the fact that so many individuals and groups were coming to Ka'upulehu had begun to generate its own momentum, which frequently led to more press coverage, speaking invitations, and requests for tours.

Similarly, our research programs were achieving considerable success on their own and contributing to and benefiting from the group's restoration and outreach efforts. My colleagues and I were doing all the things good scientists were supposed to do—landing prestigious grants, publishing papers in high-quality peer-reviewed journals (including *Conservation Biology*, which had accepted the revised version of our originally rejected manuscript), presenting seminars to academic and lay audiences. And the scope of our collective research had expanded to investigate an ever-broadening spectrum of topics that now ranged from the potential benefits of inoculating native plants with mycorrhizal fungi to the novel employment of stable isotopes to detect whether and to what extent fountain grass was altering fundamental ecosystem processes such as water cycling and carbon flow.

Interestingly, the implementation of all this science at Kaʻupulehu had also increased the value of this site and enhanced our restoration and outreach programs. Even when people had no idea what we were doing, they seemed to enjoy seeing "the scientists" in action in the field and appeared to appreciate the fact that we were spending so much time and effort learning more about this forest. Many were especially impressed by our expensive and unintelligible-looking equipment. Consequently, some of our experiments ended up being prominent and valuable components of the standard Kaʻupulehu tour.

As Susan and I discussed the joys and frustrations of doing science at Kaʻupulehu, it occurred to me that the relationship between scientists such as ourselves and conservation-oriented community groups such as the working group had many of the same dynamics of mutualistic associations in nature. People often think of these interactions as straightforward examples of idyllic win-win situations. For example, when an insect pollinates a flower, the insect gets food (nectar, pollen, or both), the plant gets sex (i.e., the insect deposits pollen on the flower's stigma, which then germinates and transfers its sperm to the flower's ovules), and everyone is happy. But the truth is that there is no referee in nature to ensure fair play in these kinds of deceptively simple mutualisms; on the contrary, there is often strong pressure for both sides to cheat if and when they can get away with it. Thus, natural selection may strongly favor a plant that happens to, say, find a way to trick some insects into delivering its pollen without having to feed them nectar, or an insect that figures out how to steal some plants' nectar without providing pollination services in return. Consequently, these "mutualistic" relationships tend to be highly complex and may shift back and forth from competitive (both sides suffer) to exploitative (one side benefits and the other side suffers) to commensalistic (neither side benefits or suffers) to partially or wholly mutualistic (both sides benefit).

The dynamics of my relationship with the working group had in many ways become analogous to these kinds of shifting ecological mutualisms. On the positive side, it was easy for me to see the many ways in which my fellow scientists and I contributed. First and perhaps most fundamentally, having professional scientists on board greatly enhanced the group's credibility and status. Second, our stature, contacts, and scientific activities helped the group get money. Sometimes we did this directly, via such actions as forming cooperative agreements with and writing nonresearch grant applications to community groups and government institutions. At other times our scientific accomplishments and ongoing research activities at Ka'upulehu indirectly helped the group raise its own money from sources that otherwise could not or would not have contributed.

Third, our professional training and our research activities helped the group develop and implement its outreach and restoration programs. Because of our educational backgrounds and research and writing skills, we often wound up taking the lead on such tasks as writing reports, producing nontechnical brochures and presentations, and developing and leading ecologically informed on-site tours and activities. Perhaps our continual efforts to employ the intellectual qualities typically required to do good science (rigor, objectivity, organization, discipline, curiosity) also helped the group value and attempt to employ a similar approach in their outreach and restoration projects. Finally, in addition to our science and its trappings becoming important site attractions in themselves, many of the physical components of our experiments (e.g., established populations of native plants and irrigation networks) often proved to be of practical value to the group's restoration efforts.

While it was often all too easy to list the costs of being an active member of the working group—the considerable time commitments, the physical and conceptual restrictions on my research programs, the additional layers of bureaucracy and interpersonal dramas—in retrospect I realized that it was also too easy to discount the many benefits I derived as well. For example, my association with the group similarly enhanced my own credibility and fund-raising success because some individuals and organizations greatly value and preferentially support these kinds of collaborations. My research programs likewise benefited from the group's considerable logistic, financial, and political support; without these kinds of benefits, there was simply no way I could have attempted something as ambitious as the Big Experiment.

Looking back, I realized that I also derived three less tangible yet personally valuable benefits from my association with the working group: the gratification that at least some of my work was "making a difference" (which was why I decided to become a scientist in the first place), the intellectual and spiritual enrichment of interacting with such an incredibly diverse community, and the moral support and camaraderie of working with people who similarly believed that highly degraded ecosystems such as tropical dry forests were worth fighting for despite the long odds.

Moreover, when things were running smoothly, the distinctions between such categories as research, outreach, and restoration became arbitrary and irrelevant. For instance, after I spent a day with a local college class doing some much-needed weeding and data collection at Ka'upulehu, I realized I had simultaneously accomplished all three of these objectives. And since both the working group and my scientific colleagues and I needed many of the same things in the field (firebreaks and access roads, ungulate fences, water and irrigation systems, native plants and seeds, volunteer coordination and labor), working together on these kinds of tasks often truly was a win-win situation.

And yet, again much as with complex mutualisms in nature, my relationship with the working group was also highly dynamic, and at times the distribution of costs and benefits could be, or at least feel, quite asymmetrical. During times when one or both of us desperately needed more of some limited resource, our relationship could quickly slide toward the competitive end of the spectrum. ("The money from that *research* grant was supposed to be for doing science, not paying for your brochures!"; "I am too busy to write another one of your reports!" Or, conversely: "The money from that *community* grant was supposed to be for outreach and restoration, not paying for more water for your experiment!"; "The technicians and volunteers can't devote any more time to your experiments because they desperately need to catch up with the planting and weeding!") Of course, these kinds of tensions were by no means limited to science-practice conflicts; at times other individuals and factions within the larger group similarly believed that they were giving too much and receiving too little.

Nevertheless, our group had somehow managed to continue fulfilling its mission. Indeed, I was often surprised (and a bit horrified) when people told me that this group was generally considered one of the more effective and harmonious such collaborations in the entire state. And despite my frustrations, I deeply believed in this community-driven restoration model and wanted to remain an active member of this collaborative team. So how, I wondered, can we change things to maximize all the good and minimize the bad?

From its conception, the group had always been a loose consortium of people and institutions that had coalesced around their shared interest of preserving and restoring native dry forests. The informal, consensus-driven, voluntary nature of this association was probably a major advantage in the beginning—anyone could simply show up and participate without having to make a formal commitment. But as the group's size, scope, and ambition grew, perhaps this informal structure was no longer appropriate. Nearly fifteen years after its creation, had the time finally come to become an official entity, such as a tax-exempt nonprofit organization? As much as I loathed the thought of yet more bureaucracy in my life, I saw the advantages of transforming the working group into an institution that could, for example, directly accept grants and donations (rather than continuing to "launder" them through other organizations); create more formal and efficient lines of communication, responsibility, and authority; and perhaps more effectively bridge our widening science-practice gap.

But as with so many other aspects of the group's operation, even if we all agreed to institutionalize, the big question was always "Who will take the lead and do the work to make this happen?" I sure didn't want to, nor did Susan; we both were already far too busy. Moreoever, the steps involved in such a task went far beyond our expertise, interests, and (as our bosses would emphatically point out) job descriptions. Yet this was more or less true for all the other members of the group. Thus, even though it might be to our collective advantage to make such changes, our default mode always seemed to be to take the path of least resistance.

What else could we do? Try folding our group into one of the Big Island's existing nonprofit environmental organizations? Hire some independent outside person or agency to run the show? Bring in a professional facilitator or shrink to help improve our interpersonal dynamics, rebuild our trust in one another, and refocus our work around the vision that initially brought us together? None of these ideas seemed very appealing or feasible. From what I could tell, most of the island's environmentally oriented nonprofits already had more than their share of internal drama. I also knew from previous unpleasant experiences that bringing in an outside person or entity to run the group or help us operate more harmoniously could itself prove to be yet another time-consuming and contentious process. Moreover, it might not necessarily make things better and could very well make them worse.

As Susan and I approached Hannah Springer and Michael Tomich's house, I realized that there probably was no easy answer or quick fix to the group's problems and that a certain amount of conflict and tension might just be intrinsic to broad coalitions such as ours. Once again, I wished I had more training in and experience with these fundamentally important aspects of the science and practice of restoration ecology and conservation biology.

When we drove past the section of the Mamalahoa Highway that separated our six- and seventy-acre exclosures, I looked longingly out the window, thought about all the fieldwork I urgently needed to do, and momentarily considered skipping our three-hour meeting. But after mulling it over, I resolved to redouble my efforts to focus on science, avoid getting sucked into the group's internal dramas, and for now "just say no" to any additional time-consuming activities that were extraneous to my research program. Maybe, I told myself, it's time to try being a dispassionate scientist who does his work and leaves it to others to decide whether and how to apply his results.

When my time on the group's agenda came up, I passed around a brief nontechnical report I had prepared to summarize the most recent results of the Big Experiment. I had planned to argue that since this research clearly demonstrated that bulldozing was our most effective and efficient technique for simultaneously controlling fountain grass and reestablishing native dry forest plant populations, we should try restoring another degraded section of the seventy-acre exclosure with dozers or encourage others to implement dozer-driven restoration projects and support their efforts. This time, however, I simply presented my information, made sure it was intelligible to the rest of the group, and shut up.

I gleaned two major insights from the subsequent discussion of my report. First, my results didn't sway anyone who had been strongly opposed to using bulldozers in the first place. Second, like much of the public in general, which tends to respect and value "science" but isn't very interested in the technical details of the research itself, most people around the table did not want to focus on the specific results of this or any other experiment. Instead, they based their perspectives and arguments primarily on their preexisting emotions, their personal experiences, and secondhand information. Consequently, the group's debate largely revolved around the same things it had when I first proposed experimenting with bulldozers at Ka'upulehu.

While it was tempting to attribute this outcome to the fact that Susan and I were the only scientists there, I had seen the same thing happen in the debates of "hard-nosed" professional researchers. In fact, observing how quickly the group's conversation spiraled away from my experimental data reminded me of the first time I led a discussion of a technical scientific paper in graduate school. Before I had even finished my overview of that experiment's results and conclusions, a seasoned professor jumped in with, to put it mildly, some strongly worded statements about the ignorance and incompetence of that paper's authors. His comments catalyzed a heated debate (based largely on emotions, personal experiences, and secondhand information) that quickly veered away from the topics addressed by the research I had presented. As the debate unfolded and my various attempts to refocus it were rebuffed, it finally dawned on me that few if any of the scientists who were doing most of the talking had actually read the paper.

Before long, Andie Beck, our facilitator, broke in and urged us to wrap it up and move on because, as always, we needed to address several other pressing topics before we adjourned. Predictably, the group decided that before taking any action it would be best to let the Big Experiment run its course and see how well these initial results held up over time. "Okay," Andie said, "let's talk about the possibility of hosting commercial tours of Ka'upulehu. Several groups have expressed an interest in bringing people here, and we were supposed to get back to them months ago. We need to think about whether we want to do this, how much we should charge if we do, liability issues . . ."

For a while, my decision to focus strictly on research seemed to make everyone happy, including me. My supervisors were happy with my increased scientific productivity and "rapidly growing maturity." My colleagues were happy because once I stopped trying to bend our science toward solving applied problems and achieving conservation goals, our research projects became more rigorous and academically interesting. The people on the ground were happy that I was no longer meddling in their affairs; in fact, at Ka'upulehu I successfully lobbied for separate designated areas for science and restoration, which seemed to help us all literally and figuratively stay out of one another's way. And I was happy because for once I didn't feel so schizophrenic—I knew what I was doing and why I was doing it, and I suffered no delusions about the practical value or future applicability of my research.

Unfortunately, this bliss didn't last. Within a few months, three little cracks in my "science-only" armor developed and began spreading toward my heart. The first crack was that while I might have finally decoupled the academic and practical values of my research, the external world kept pushing me to mix the two entities back together. Everywhere I turned in my professional life—the papers and grant applications I wrote or reviewed, the talks I presented or heard, the workshops and outreach events

I participated in or attended—it seemed that we all were supposed to devote at least some time to arguing how practically valuable our research programs and our good science in general were or would soon be. Apparently we were also supposed to urge the scientific and practitioner communities to communicate more effectively and work together to solve our pressing ecological problems before it was too late.

Second, since I had distanced myself from applied fieldwork, I was no longer willing or able to serve as a bridge between the scientists and bureaucrats at the conference table and the practitioners on the ground. This was particularly true at Ka'upulehu, where the consequences of my withdrawal from that role became increasingly difficult for me to ignore. For example, I knew the guys who were doing most of the plant propagation and outplanting in the seventy-acre exclosure reasonably well. For the most part, they were smart, hardworking, dedicated people who truly cared about native dry forests and wanted to see them restored. But they came from and lived in another culture from the one us *haoles* did. They hadn't had much formal education; they didn't know or care about the more abstract, intellectual issues surrounding our work; they didn't attend the working group meetings and were largely unaware of our debates and decisions. Not surprisingly, as I had observed with the outplantings at the end of Don and Alan's trail, they increasingly decided to just do what they wanted to do.

Like most other mission-oriented practitioners, what they wanted to do in this case was the "real stuff": kill weeds, dig holes, plant trees. As far as they were concerned, scientific research in general, and time-consuming, tedious tasks such as labeling and measuring plants, and creating and maintaining computer databases in particular, were neither important nor within their job descriptions. With Susan and I frantically trying to stay on top of our own mountains of data, all of that "data kind of stuff" at Ka'upulehu (as was happening in many of the other restoration programs I observed) slowly but surely began to unravel.

Consequently, while I was pleased with the speed and scale at which we were accomplishing such tasks as killing fountain grass and planting native species, I was growing increasingly concerned about the haphazard and uncoordinated design and implementation of our restoration program as a whole. For example, in addition to all those undocumented and neglected outplants, the placement and composition of many of the plantings themselves had become ever more driven by crisis and convenience rather than by proactive planning and incorporation of the fundamental principles of restoration ecology and conservation biology.

The third and final crack in my science-only armor was that I was start-ing to lose my motivation for performing research at places such as Ka'u-pulehu. If, as I'd been trying to convince myself, the point of this science was to improve our ecological knowledge of highly degraded ecosystems, then why continue to work here and collaborate with the working group? There was no shortage of degraded ecosystems that didn't have the ex-treme logistic, philosophical, political, and social challenges of this study system and community group. Why not just do science elsewhere and vol-unteer with conservation-oriented organizations and projects in my free time?

Yet as I tried to visualize throwing myself into a brand-new research program focused on more esoteric, academic questions, I realized that my heart just wouldn't be in it. I knew that many other scientists managed to effectively separate their professional research from their personal conser-vation interests. While I respected their decisions and valued the more ba-sic knowledge and insights their science yielded, this approach wouldn't work for me, in part because I now found the broad, interdisciplinary chal-lenges of applied ecological research more compelling than the narrower, more specialized challenges of academic ecology.

Similarly, while I had always been aware of the importance of various philosophical issues to the theory and practice of both academic and ap-plied ecology, in my zeal to get my "real" work done, I had initially been content to leave all that "soft stuff" to the philosophers and social scien-tists. But years of struggling to navigate the often tortured maze of philo-sophical paradoxes and ethical dilemmas I encountered while trying to marry the science of restoration ecology to the practice of ecological resto-ration had changed my perspective. In fact, I now often found these kinds of problems at least as interesting and important as the more technical and ecological challenges.

At Ka'upulehu in particular, philosophical issues had become as ubiq-uitous and dominant as fountain grass. They also seemed to permeate an increasing portion of the working group's discussions and decision-making process. The following three case studies provide a sample of the diversity, complexity, and importance of the kinds of philosophical conundrums we regularly had to address:

1. Black cypress (*Callitris endlicheri*). In the 1950s, territorial foresters planted several different nonnative tree species within the six-acre Ka'upulehu exclosure. One of these specimens, a black cypress,

eventually grew taller than all the other trees in the area and thus became an important regional landmark. Some local members of the working group felt that because this tree was historically important and ecologically harmless (it had failed to regenerate since it was planted), killing it would be unnecessary, insensitive, and counter to our larger mission. Other, mostly nonlocal members argued that this tree had no place in a native dry forest preserve and should therefore be removed.

2. *Kukui*, or candlenut tree (*Aleurites moluccana*). Although this species is the official state tree of Hawai'i, it is actually an alien species that was deliberately brought to the islands by the early Polynesians. Naturalized *kukui* trees now thrive in many different habitats across the archipelago (including parts of North Kona but not within the Ka'upulehu exclosures), but they tend to form dense, dominant stands only in riparian areas. Because of its cultural importance, beauty, and ability to grow relatively quickly and provide substantial shade, some members of the group wanted to explore the efficacy of planting *kukui* in a few highly degraded areas of the seventy-acre exclosure to create favorable nurse environments for the eventual establishment of native dry forest species. Some members were dead set against planting *kukui* or any other alien species under any circumstance; others were willing to try *kukui* and other species originally brought over by the early Polynesians; and a few were willing to experiment with any promising noninvasive species regardless of its geographic origin.

3. Tree tobacco (*Nicotiana glauca*). The beautiful Blackburn's sphinx moth (*Manduca blackburni*) is Hawai'i's only federally endangered insect. As with many other native insects, the rarity of this species is due in part to the rarity of its key host plant, which in this case is 'aiea (*Nothocestrum breviflorum*), a native dry forest canopy tree that is also federally endangered. Although there are a few mature 'aiea trees scattered across the Ka'upulehu exclosure, we had never observed a sphinx moth on them. (Given our limited expertise and search efforts, it was possible that this moth had occasionally used some of those trees as a host.) However, we and others had observed this moth at Ka'upulehu on tree tobacco, which is in the same family as 'aiea (the Nightshade family) but is a fast-growing alien tree that has infested portions of North Kona and other arid regions of Hawai'i. To support at least the short-term survival of the

sphinx moth, should we leave Ka'upulehu's relatively small number of tree tobacco plants alone, or should we eradicate this noxious weed before it is too late?

Even if I had wanted to leave the resolution of these questions to the
other members of the working group and just focus on my science, I still
would have had to wrestle with a stream of my own vexing philosophical
issues. In addition to the big, overarching questions ("What exactly am I
trying to accomplish here, and why?") and ethical challenges ("Should I
destructively sample all those experimental endangered plants to maximize my statistical power or spare some for restoration purposes?"), I often
struggled with more mundane but potentially critical scientific methodological issues. For instance, around the time we were trying to decide
what to do about those three alien tree species, an endemic native vine
in the Gourd family (*Sicyos lasiocephalus*) appeared out of nowhere,
climbed up the poles supporting one of the strips of shade cloth in the Big
Experiment, and eventually produced a dense mat of vegetation that significantly reduced the light in several of the plots along that strip. Should I
cut it down, or should I leave it alone because it is native?

For additional guidance, I tried reading the most relevant philosophical and social science literature I could find, and I even discussed specific
cases with a few experts in these fields and attended some of their conferences. While I learned a lot from these intellectual adventures, they did
not help me solve my philosophical problems. I eventually concluded that
in addition to the science-practice gap in applied conservation, there appeared to be a substantial gap between the work of the "applied" environmental philosophers and social scientists and the complex issues we were
trying to address in the real world.

Despite our best efforts, we also collectively failed to solve our more
practitioner-oriented philosophical issues in any rational, consistent, or
coherent manner. I finally came to accept the fact that we were going to
have to resolve these dilemmas on a case-by-case basis because they were
just too idiosyncratic, subjective, emotional, and value driven to be addressed by overarching academic theories and general principles.

Similarly, was tackling the unique ecological and technical challenges
of each degraded ecosystem on a case-by-case basis also the best we could
do in restoration ecology? Why is there such a gap between the science
and the practice of these applied, environmentally oriented disciplines? Is
this gap intractable, or are there feasible steps we can and should take to
help bridge it and develop more mutually beneficial relationships be-

tween scientists and practitioners? Can applied scientists modify their research paradigms and methodologies so that they produce more robust and practically valuable results and insights for the practitioners in these disciplines? These are the kinds of general questions I explore in part 2 of this book.

Toward a More Perfect Union

Chapter 7

The Science-Practice Gap

After struggling to support ecological restoration with my own research programs, I became increasingly interested in the dynamics of other restoration projects and the extent to which science and scientists were or were not providing them with practically valuable information. To investigate these questions, I explored the ecological and institutional frameworks of other restoration programs across the Hawaiian Islands. I also interviewed people involved in these programs and in other areas and disciplines related to the science and practice of restoration ecology, such as academia, conservation biology, environmental education, and resource management. Here I explore the nature of this science-practice gap; in the next chapter I offer some solutions for bridging the gap.

It was both gratifying and sobering to repeatedly find that my struggles were by no means unique. Indeed, while I did discover some encouraging examples of good relationships between scientists and practitioners, I found that these two communities were, on the whole, deeply polarized and disconnected from each other. When I asked people why they thought this relationship was so poor and unproductive, many who were not members of the scientific and practitioner communities highlighted the different backgrounds and cultures of most of the individuals in these two groups. One person who worked for a nonprofit environmental organization put it this way: "A lot of the friction comes from the fact that scientists have more education, greater prestige, better pay, and more freedom than practitioners do. They're out there having fun, doing whatever they want to do, while the practitioners are doing hard, often tedious physical work. No wonder some of them have a chip on their shoulder!"

Many people in both the scientific and practitioner communities stressed how little they felt the "other side" understood and appreciated what they did. Perhaps the most common overarching theme I heard in the comments of practitioners and other nonscientists was that because most scientists do not understand the complexity and difficulty of actually doing restoration ecology and conservation biology, they tend to grossly overestimate the relevance and value of their relatively narrow and abstract research programs.

Many scientists in turn pointed out that because most practitioners do not appreciate the complexity and difficulty of "real science," they do not understand why scientists can't perform the kinds of relatively simple, commonsense studies they request. Nevertheless, a substantial number of the researchers I interviewed claimed, and I think sincerely believed, that their own and other forms of "real science" were of far more practical value than the practitioners realized. One scientist said, "Practitioners don't appreciate how much of their knowledge and tools come from science. It's easy for them to take the fruits of science for granted and say 'We already know all that' when the fact is that until scientists did the research, they didn't know or understand it at all! Good science can also help them prioritize and focus their management activities and reveal the underlying complexities of things. If they think they already know everything they need to know, they are ignorant, and that's their problem, not mine."

Interestingly, however, when I asked scientists for particular projects and specific issues in which this basic science had proven to be practically valuable, many stated that although they could not personally think of any examples, they knew they were out there. On the relatively rare occasions when someone did mention a real-world example of the practical relevance of rigorous science in applied restoration ecology or conservation biology, on further inspection I always found a far more complicated and less convincing story.

One of the best examples of this scenario was the US Army's massive preservation and restoration program on its 3,100-acre Makua Military Reservation on the island of O'ahu. Perhaps because I knew several practitioners who had been working on this project for years, and I had often heard them vent their frustrations over such things as its "snail's pace" and "exorbitant, misguided costs," I was somewhat surprised when several scientists independently cited Makua as being one of the most convincing demonstrations of the practical value of basic research and the power of "science-driven restoration." One senior scientist and passionate conservationist who tends to be highly critical of restoration programs in general put it this way: "I've been intimately involved with this project from the start, and I can assure you it's one of the best, if not the best, in the entire state. For once, we proceeded slowly and methodically and conscientiously, and based our decisions on sound, rigorous science. If you want to see an example of 'restoration done right,' go check out Makua."

I did. Unfortunately, while I was able to interview several people directly involved with the design and implementation of this program, because of Makua's complicated logistics and security issues, as well as my own time constraints, I was unable to get any closer to the actual fieldwork than the office and administration buildings just inside the entrance to the base.

Like many such "species-driven" projects, the Makua restoration program was mandated by law. As a federal agency, the US Army is required under the Endangered Species Act to consult with the US Fish and Wildlife Service on the potential negative impacts of its actions on any threatened or endangered species. Although the army has been conducting live fire exercises at Makua since World War II, after the USFWS classified a whole slew of Hawaiian plants as federally endangered in the 1990s, it suddenly found itself with forty-one endangered plants and one endangered snail in the line of fire. After some lengthy legal battles and contentious negotiations, they ultimately decided that twenty-seven of these plants and the snail required "additional stabilization actions." The army was thus ordered to put together an expert implementation team to craft a specific plan for stabilizing each of these twenty-eight species, a task that wound up requiring three thick volumes to complete.

"We ended up having 125 tough, grueling, all-day meetings," one member of the Makua implementation team told me, shaking his head. "Do you know how much it cost to fly all of us over to O'ahu and house and feed everyone? And the plan we came up with was going to cost $600 million! All to try and save a bunch of mostly hopeless plants growing in weed patches in heavily degraded ecosystems on a live fire military base? Meanwhile, the rest of the conservation community is struggling to save the last few intact, relatively pristine places on a shoestring budget . . . I just increasingly felt that the whole thing was unconscionable."

Several people also told me that despite all this time and effort, they did not believe that the design and implementation of their management plan were actually "science driven." For example, when I asked another team member, who had many years of experience in Hawai'i, about this issue, he laughed. "I read more basic ecology and conservation literature for this project than I had in all my previous years combined," he said. "It's true that we really tried to stick to the science to guide us through this project at first, but the truth is that it just proved to be too nebulous, incomplete, conflicting, and inapplicable to Hawai'i and our work at Makua. So in the end, although some people don't like to admit it, it really all came down to our gut instincts."

Another member of the team with extensive field experience on O'ahu stressed what in her opinion was an ongoing major gap between the conference table discussions and plans and the on-the-ground realities. "We tried to quantify everything at first and base our management actions on a formal, objective point system. We wanted to establish at least three stable populations of at least fifty mature individuals per species and integrate each of these populations into the larger management units. We argued a lot over whether that was a reasonable and defensible management objective, but in the end a lot of those arguments were moot because it often just wasn't possible to do what we finally agreed to do in the field anyway. We don't have a lot of flat areas and access roads at Makau—what we've mostly got are cliffs! We took the whole team out in the field three times to try to show them what we were talking about, but somehow, back in the office, people just forgot what they saw.

"Not surprisingly, the cost of the plan we came up with was staggering. If the whole thing had gone through as we originally envisioned it, we would have had to hire 150 people, and it would have cost more than the army's entire national environmental budget to implement. That's partly because there were so many endangered species out there, partly because we were overly ambitious, and partly because it just costs a lot more money to do things under those conditions than people realized. So we had to go back to the table and argue about it all over again."

"A lot of our plan ended up requiring us to work in some really degraded areas because we were so desperate to find more habitat to 'stabilize' each endangered species," another member of the implementation team recalled. "We argued about the relative merits of in situ versus ex situ sites, but like everything else, we never really resolved this issue because I think it's really just irresolvable—you can find data and literature to support whichever way you want to go. But the thing that really scares me is how gray the Endangered Species Act is—because it's so vague, its 'official' interpretation is always changing with the political winds. And since Congress never reauthorized this act, they could nix our budget at any time. We're down to $4 million a year now, which isn't nearly enough, but no one knows how long we'll even be able to hang on to that much funding because the army has never actually committed itself to implementing our entire plan."

As has been the case in virtually all of the applied restoration and conservation programs I have seen in Hawai'i, I came away with great respect and appreciation for the dedication, skill, and work ethic of the people on the ground at Makua. Many were frustrated by the logistic, ecological, and bureaucratic constraints surrounding their project, especially the legal requirement to focus on individual species as required by the Endangered Species Act, rather than whole communities and ecosystems. (Many others within and beyond the Hawaiian conservation community expressed similar frustrations.) Nevertheless, they were determined to make the best of the hand they were dealt and find a way to get good things done.

Most members of the Makua implementation team with whom I spoke also wound up appreciating, though not necessarily agreeing with, one another's point of view. However, few believed that the long and arduous process they went through to create their management plan was worth it in the end or that the rigor of their planning process would necessarily translate into higher-quality management actions.

One senior member of the team diplomatically summed up their collective experiences at Makua this way: "We all learned a lot from absorbing so many different perspectives around the table—the theoretical people, the systems people, the single-species people . . . But to date the results have been so patchy, it's hard to know how successful this plan will really be, and it's hard to come up with any overarching lessons from all the energy we devoted to this project. When we finally finished our work and went our separate ways, I think we all had a great deal of respect for each other, but there was also a deep division between the few people who thought it had been a good, productive process and the rest of us, who thought it had been an ugly, painful, and wasteful affair."

Another common complaint I heard from nonscientists was the tendency for scientists to dominate their collaborations and usurp the lion's share of their limited resources. Such practices obviously widen the science-practice gap and foment resentment of science and scientists. A field technician who had worked at Makua and in many other restoration projects told me, with more than a hint of disgust in his voice, "Of course, once again, our whole purpose [at Makua], which was supposed to be to preserve and restore endangered species, got hijacked for esoteric, academic pursuits: 'Let's learn a little more about this along the way, let's learn some more about that . . .'"

At the same time, however, many scientists and practitioners stressed that most restoration programs desperately need *more* scientific guidance. Several people responsible for designing or supervising on-the-ground projects apologetically told me that although they wished their work was more science based, they simply were too busy, untrained, or underfunded to do it themselves. Because they had also been unable to get outside scientists to work at their sites or help them design and implement their programs, they did the best they could on their own. A good representative example of this situation is the Kōke'e Resource Conservation Program (KRCP).

KRCP is a volunteer-based alien species control program dedicated to the preservation and restoration of native forests on Kaua'i. This organization works in the heavily visited Kōke'e State Park, Waimea Canyon State Park, and Nā Pali Coast State Wilderness Park on the island's rugged and remote northwestern side. Collectively, these parks encompass more than 12,000 acres and provide critical habitat for twenty-six federally endangered plant species. Much as with the "special ecological areas" in Hawai'i Volcanoes National Park on the Big Island, KRCP concentrates its efforts in the relatively intact and accessible areas of these three state parks.

"We really try to focus on those areas because so many of our local folks and outside visitors go there," Katie Cassel, the program's director, told me while she and I and Ellen Coulombe, KRCP's volunteer coordinator, were chopping our way through a dense patch of invasive ginger with some volunteers from the mainland. "We've really turned back the clock on the weeds in some selected areas, but of course it requires infinite follow-up and persistence. Usually we have to stay on top of them for the first two years, until the natives start reappearing; then the balance starts shifting and we can get away with doing less and less. Some places that were 100 percent weeds five years ago are now dominated by a diverse collection of native species!

"We've learned to go after the ecosystem-destroying weeds—the guavas and gingers—and not waste our time on the little stuff," Katie explained. "And we've gotten better at helping the volunteers stay focused and be more efficient by working contiguous areas. But at the same time, we're always scrambling just keeping up with the weeds, training and motivating the volunteers, and looking for more funding. I wish we had a really good scientific monitoring program to better assess the overall effectiveness of what we're doing, but we've never had the time or money or staff to do that. And it's all so patchy anyway; everything keeps changing, and lots of strategies that sounded good on paper just haven't worked out."

"The volunteers are wonderful," Ellen added. "A lot of them get caught up in this work and will do things like adopt a particular area and come back and weed it on their own, which helps us out tremendously. But they also require a huge amount of our time—the logistics of recruiting them, getting them up here and to the different sites and back, explaining our overall mission, teaching them the plants and how to handle the tools and herbicides safely . . . We've learned that a lot of our success depends on assessing each new group of volunteers and choosing the right battle for them that will provide the right balance of challenge and satisfaction."

"Then of course there's the challenge of the weeds themselves," Katie said. "We have to carry so many different tools and herbicides to kill all the different ones we encounter. We try to stay focused and disciplined, but there are always new ones coming in, or new populations of existing weeds springing up in remote, pristine areas, and we always want to try and control them before they establish and spread and become unmanageable. And everything ultimately depends on the ebb and flow of our funding and our supply of volunteers. We're always short on staff and facilities and supplies. If we had twice as much as we do now, we could probably get at least twice as much done. It would also help tremendously if the state would ever commit to fencing some of these lands, but due to local politics and the hunting lobby, that hasn't yet happened, so all of our work takes place in areas that are still open to the pigs and deer."

Despite all these and many other challenges and limitations, they have managed to accomplish a heroic amount of brute-force, just-do-it work. For instance, according to their records, between 1998 and the summer of 2005, KRCP's 14,985 volunteers put in 8,166 person-days and killed 6,246,374 weeds over a 4,139-acre area!

When I asked Katie and Ellen about the relevance of science to their work, they both stressed its importance and how much they wished they could find a way to get more of it. Yet when I asked for more specific details, I found that the science they had incorporated into their work and wanted more of consisted of informal, localized projects in which someone such as a US Department of Agriculture (USDA) extension agent had investigated a narrow, applied topic such as the efficacy of three different herbicides on two problematic weeds.

When I asked whether any more basic, academic science had ever been practically relevant to their work, they stared at me blankly. I soon discovered that, like most of the other practitioners I interviewed, they had almost no exposure to that world and had never heard of any of the "applied" science journals I mentioned that were ostensibly designed to help guide and inform their work.

"We'd love to have more information, and we're always open to anything that might help, but I guess we just never run into those kinds of scientists out here," Katie said. "We do try to send people to the Hawai'i Conservation Conference when we can, to keep up on the latest stuff, but I don't know how relevant any of that has been so far. It's great to hear what's happening with, say, the ginger in Hawai'i Volcanoes National Park on the Big Island, but the truth is that every situation seems to be different, and there are so many variables. Mostly what we do is more intuitive stuff, although we do try to be scientific about it."

"We do cite one of those papers that talks about how alien species can alter ecosystems in our grant applications," Ellen said earnestly, "and maybe that has helped us obtain some funding? Of course, I'd like to think that anyone who has spent any time in the forests up here and knows the difference between a native and an alien plant would already know that!"

After I investigated these and many other applied programs, and reflected on my own work and conversations with scientists and practitioners, it became increasingly clear to me that these two communities have very different ideas of what science is and how it should be done. These different conceptions are important because they are yet another significant source of tension and conflict.

For instance, practitioners tend to think of science as any careful, systematic approach that involves recording data and making careful observations. Thus, they might "scientifically" investigate, say, the effect of a new plant fertilizer by applying it to a bunch of seedlings in the field and then perhaps documenting their results by taking photographs or jotting down qualitative notes.

On the other hand, academically trained researchers (and their corresponding scientific journals, funding agencies, and employers) have a much more formal and narrow perspective of science (hypothesis formation and testing, replication, quantitative data collection and analysis, rigorous and sophisticated statistics, etc.). If these scientists were interested in the effects of this fertilizer, they most likely would "do it right" by setting up an experiment with a large number of seedlings and randomly selecting half of them for the fertilizer treatment and half as unfertilized controls. They would then strive to control, standardize, and quantify as many of the relevant variables as possible—the genetic background and weight of the seeds used to grow the plants; the amount of light, water, and fertilizer received by each seedling; initial seedling size and spacing—as well as meticulously measure and formally analyze a series of predefined variables at the end of the experiment, such as plant height and width, leaf area, biomass, and water-use efficiency.

Consequently, formally trained scientists may consider the "research" typically performed by practitioners as naïve, pseudoscientific, and uninterpretable and may criticize these protocols and the practice of restoration in general as undisciplined and overly reliant on "uninformed gut feeling decisions." Conversely, practitioners may view the experiments and procedures of formal scientific research as too slow, expensive, and reductionist and may retort that the scientists, whose research is ostensibly designed to help them, have little comprehension of the real-world practice of restoration.

Obviously, these very different paradigms of "real" science can complicate communication between scientists and practitioners and inhibit or even sabotage their collaborations. Perhaps because I have had multiple opportunities to perform both "formal" and "practitioner" science, I can appreciate the relative value and appropriateness of both schools of thought. Much to the surprise of many entrenched members of these two communities, both of these approaches also match at least one of the technical, dictionary definitions of science. For instance, the practitioner's methodology conforms well to the *Oxford English Dictionary*'s "2b" definition of science: "Skillful technique, especially in a practical or sporting activity," and the more formal, academic approach is well described by *Oxford*'s "3c" definition: "An activity or discipline concerned with theory rather than method, or requiring the systematic application of principles rather than relying on traditional rules, intuition, and acquired skill."

Both approaches also fall under the same, more open-ended definitions and paradigms of science typically employed by social scientists. For example, one eminent ecological historian wrote, "Science is an ongoing negotiation with nonhuman nature for what counts as reality. Scientists socially construct nature, representing it differently in different historical epochs." Thus, depending on one's chosen definition and philosophical position along this continuum, it is possible to legitimately consider, say, green-thumb plant propagators, insightful indigenous peoples, and skilled farmers as being either among the best ecological restoration scientists or entirely outside of and irrelevant to this discipline.

Given this diversity of formal definitions of and approaches to science, it is not surprising that even within the same professional communities and subdisciplines, individuals often have radically different definitions of good and bad science. This fact was dramatically illustrated to me when I served on a USDA Biology of Weedy and Invasive Species panel charged with ranking the merits of over a hundred grant applications. After laboriously and at times contentiously whittling down our "must fund" list to about a dozen proposals, we were suddenly informed that because of unforeseen budgetary shortfalls, the USDA would be able to fund only a few of the grants submitted to our panel. We were thus instructed to further shorten our list to the few proposals that represented the "very best science."

The ensuing debate illustrated just how subjective the concept of good science is. In our case, did it mean research that would most likely produce the greatest contributions to our academic, theoretical frameworks for modeling the biology of invasive species? Were clever and sophisticated proposals better than more creative and simple ones? Were high-risk, high-potential experiments better than surefire but less exciting ones? . . .

As we agonized over such questions, it became clear that the only panel members who considered practical relevance to be an important attribute of good ecological science (especially when the subject was weedy and invasive species) were those of us who had at least some applied, on-the-ground experience. Yet we had to concede that the few proposals on our short list that were more oriented toward addressing real-world problems were narrower in intellectual breadth, less rigorous (fewer replications, less tightly controlled variables, more collaborations with practitioners, etc.), and riskier than those that were focused on more abstract and theoretical issues. Several of the more academically oriented researchers on the panel also made the familiar argument that the grant applications in the latter group could lead to the development of conceptual models and tools that would eventually be valuable to the broader practitioner and conservation communities. Given that our panel was dominated by the academics, no one was surprised when we ultimately chose to fund the most basic, theoretically oriented proposals. But when the few panel members more oriented toward applied science told me how frustrated they were by this "typical decision," I realized that the science-practice gap can exist even within a group composed solely of scientists!

Obviously, for a variety of reasons, science in all its many forms and styles will continue to be critically important to humanity and to the worlds of restoration ecology and conservation biology. However, at least in the case of ecological restoration, despite many earnest claims to the contrary, I have consistently found that the direct practical value of bringing formally trained, rigorous scientists into complex real-world projects is often marginal at best. However, as I experienced at places such as the Ka'upulehu Dry Forest Preserve and repeatedly observed in other restoration programs, the sympatric implementation of formal science and applied restoration can result in significant, mutually beneficial synergies that in turn can facilitate more and better science and more and better restoration. Thus, incorporating research scientists into restoration programs often does effectively support such programs, even when these scientists and their brand of science have little or no direct connection to the technical development and implementation of the applied work itself.

Yet if the practical value of formal science is largely indirect, do applied restoration programs necessarily have to include formally trained scientists or subscribe to the larger paradigm of Western academic, reductionist science? Most scientists would unequivocally say yes and argue that any such program that is not at least partially informed and guided by this kind of science is not legitimate restoration. But as I discovered in my interviews and related literature research, many people outside the natural sciences (and a few within them) would say no, or at least not always.

Such debates inevitably tap into much broader, interdisciplinary arguments over such questions as "What should and should not be considered 'real' restoration, and who gets to decide?" For example, should we view the large-scale, single-species tree plantings typically performed by commercial paper companies in the wake of their clear-cutting operations as ecological restoration or profit-driven industrial agriculture? What about some of the more ambitious projects performed by landscape architects, local community groups, or even so-called nature artists?

For both well-intentioned and self-serving reasons, various individuals and special interest groups have passionately argued for different definitions of the science of restoration ecology and the practice of ecological restoration. Given the great diversity of approaches to what many of us would consider to be ecologically valuable restoration, I personally believe the widely cited definition provided by the Society for Ecological Restoration's Science and Policy Working Group is appropriately inclusive and rightly focuses on outcomes rather than specific approaches, philosophies, or credentials: "Restoration ecology ideally provides clear concepts, models, methodologies and tools for practitioners in support of their practice."

Yet under the ironic heading of "a broader definition of restoration science," several of my colleagues within the Hawaiian scientific research community recently paraphrased this definition as "Restoration science may be defined as the process through which restoration scientists can provide restoration practitioners with the 'clear concepts . . .'" Thus, perhaps inadvertently but nonetheless tellingly, their paraphrase replaces the "restoration ecology" term in the original definition with "restoration science" and then states that the concepts and tools needed by the practitioners will be provided by "scientists" (presumably meaning only those with a PhD from a Western-style university).

This perspective, which in my experience is widely held throughout the formally trained research community, can obviously lead to conflicts and power struggles between scientists and practitioners. Indeed, even some academics have warned that the effective practice of ecological restoration may now be in danger of being subsumed by this "scientific authoritarianism" and have thus argued for a broader, more holistic approach with greater respect for other kinds of knowledge and ways of learning about and interacting with the natural world. Some have also pointed out that the relevance of this kind of formal science is often dwarfed by the overriding importance of the various aesthetic, cultural, socioeconomic, philosophical, and political components typically associated with ecological restoration in the real world.

Yet even when I temporarily put aside these concerns and arguments, which I consider to be quite valid, I still found that the formal scientific approach my colleagues and I were trained to employ and revere was not necessarily the most effective framework and methodology for quantifying and assessing complex ecological phenomena and testing hypotheses geared toward guiding on-the-ground programs. In other words, despite its many other virtues, I did not find that employing this research model enabled me to do a very good job of providing the "clear concepts, models, methodologies and tools for practitioners in support of their practice."

I believe that the major underlying reason for this disconnect is that the goals and practice of formal science (e.g., generalizable knowledge and conceptual frameworks acquired via methodical observation and experimentation) often directly conflict with those of applied ecological restoration (e.g., site-specific knowledge and timely on-the-ground solutions acquired via common sense and informal experimentation). Thus, from the perspective of the practitioners laboring to restore degraded ecosystems, many of the experimental methodologies we must employ within the paradigm of formal science (growing control plants in highly unfavorable microsites, suspending weeding and watering regimes, destructively harvesting native plants, etc.) often appear at best counterproductive and at worst downright stupid.

Similarly, the demands of this formal science tend to lead us toward constructing abstract general models and unifying theories that necessarily ignore or discount the diversity and heterogeneity of real ecosystems. Thus, metaphorically speaking, we often must try to fit the square grid of formal science to the round Earth of nature. While this approach may be a logical and powerful way to learn more about the world, it does not follow that it is necessarily the best way to help practitioners understand and manage the ecological nuances and idiosyncrasies of what is in fact a very round planet.

One of the very few practitioners I interviewed who had a strong background in formal science succinctly captured the gap between these distinct worlds: "I actually enjoy putting my resource management work into various scientific, academic frameworks. It's a very enlightening and entertaining pastime, but of course I do it tongue in cheek, because it always ends up being quite a stretch. The real world is just so much more complicated and messy than that!"

Of course, practitioners must also simultaneously contend with all the political, socioeconomic, and logistic factors that are usually so important to real-world projects but beyond the scope of formal science. Thus, we should not be surprised that this combination of the roundness of nature and the confounding complexities of the human world may often severely limit the practical relevance of formal research programs, even when they are explicitly designed to guide and facilitate on-the-ground projects. For instance, one of our experiments at Kaʻupulehu surprisingly found no consistent differences in the biomass of newly recruited native species within weeded versus nonweeded plots. Yet although we had designed this study in part to help optimize the efficacy of our weeding efforts, our actual weed management program was ultimately driven by a mixture of other concerns that were far beyond the reach of this experiment to address: Which combinations of microsite and native species could safely be left unweeded, and for how long? Would untrained volunteers be able to efficiently find native species within thick weed patches? How would a weedy understory affect our outreach program and future funding capabilities?

Finally, scientists often argue that one major benefit of, and justification for, formal, rigorous science is that it is generalizable. That is, its results, theories, and conceptual frameworks can be broadly applied to other situations that may be quite distinct from the parameters under which the original research was performed. Thus, if we don't harness this formidable and unique power of formal science, we will never amass more than superficial, haphazard, site-specific knowledge.

While this may indeed be the case for such disciplines as physics and chemistry, it remains to be seen whether it will necessarily be true for ecology and ecological restoration. For example, given the often extreme overall complexity and spatial and temporal heterogeneity of Hawaiian dry forests, it is unclear whether and to what extent our research results at Kaʻupulehu would hold true for other points in time and space even in that particular forest, let alone in other tropical dry forests within and beyond the Hawaiian Islands. Moreover, this difficulty in generalizing the results of ecological research is by no means unique to this ecosystem or to the Hawaiian Islands. For instance, in an extensive review of the response of Hawaiian ecosystems to the removal of feral pigs and goats, three senior scientists concluded that "the variables are too numerous and too uncontrollable to allow definitive cause and effect statements about responses of vegetation due entirely to feral ungulate removal." In fact, the effects of ungulate grazing on ecosystems throughout the world continue to be a source of often passionate and acrimonious debate even within the academic scientific community.

Right around the time that I was beginning to wonder whether any other scientists were noticing the disconnect between formal science and applied restoration and conservation programs, I read an essay by David Ehrenfeld, the founding editor of the prestigious journal *Conservation Biology*. "From the beginning," he wrote, "the journal was by most objective measures a roaring success . . . But occasionally I would experience a small spasm of doubt. Conservation biology was supposed to be, like medicine, a life-saving profession. Were we saving the lives of any species or ecosystems?" Later in the essay, he wondered whether

> deep down, conservation biology isn't really like medicine—perhaps we are just ordinary biologists trying to find comforting and trendy justifications for doing what we love to do anyway. This possibility was supported by quite a few of the manuscripts I received, which seemed to have little to do with actual conservation. Typically, after devoting 16 pages to the genetics or ecology of a plant or animal that happened to be rare, or that might some day become rare, the authors would tack on a depressingly predictable final paragraph that would explain how important this work could eventually be to conservation and why more research was needed.

Ehrenfeld then reported the results of his survey of all of the papers published in three successive issues of *Conservation Biology*:

For each of the 66 published articles I asked this question: is there strong indication that any actual conservation has been achieved already as a result of this work? Has the doctor made the patient better yet? The answer for all but three of the 66 articles was no. No matter how exciting and convincing 63 of those 66 papers were, no matter how painstakingly constructed their conservation arguments, the predicted conservation dividends were to be earned in the unspecified future. Why? Is conservation biology a delusion?

The more I investigated on-the-ground conservation and restoration projects, the more I came to see that Ehrenfeld was on to something. Despite the often dogmatic and defensive statements of the scientists I interviewed ("Of course science is critically important—it's the engine that drives all effective conservation and restoration projects") and the similar though more mechanical statements of the practitioners ("We use science all the time to design, implement, and assess our management actions"), I discovered a huge gap between such claims and the realities on the ground. In fact, when I probed a little deeper, in every case I found that one or more of the following five outcomes best described the real situation:

1. The practical value and relevance of the science was indirect, such as public education, political support, or funding and logistics.
2. The science came after, or was motivated by, some management action that had already occurred, as opposed to the oft-claimed opposite situation of the science preceding or motivating the management action.
3. The science was conducted before or during the management action but was largely or completely unrelated to the on-the-ground project.
4. The science improved our knowledge about some topic that was directly relevant to the management activity (e.g., mycorrhizal fungal ecology), yet the on-the-ground work utilized preestablished strategies and techniques that were independent of and unaffected by this science (e.g., adding small quantities of native field soil to the potted plants and outplanting holes, as propagators have been doing since long before modern science discovered the existence and importance of the mycorrhizal fungal communities in that soil).
5. The "science" was directly relevant and applicable, but it turned out to be the kind of informal research, designed and implemented

by practitioners, that would have been virtually impossible to fund and publish within the more formal research channels.

Much as in Ehrenfeld's survey of *Conservation Biology* papers, several of the scientists I interviewed similarly stressed the *potential* of formal research to change the way practitioners think, and how important and practically valuable it *could* be to their work at some undisclosed future date. While I do not know whether or not this will eventually happen, I can say that to date, despite all my earnest and extensive searching, *I have been unable to find a single clear example in which formal scientific research has been, or is, of direct and practical relevance and value to an on-the-ground restoration program in Hawai'i.* Given many practitioners' often desperate urgency to solve today's problems, such repeated but unsubstantiated assertions of the past and future practical value of formal restoration ecology and conservation biology science have understandably fueled their impatience with, and in some cases their ultimate rejection of, the relevance and utility of these disciplines.

My motivation for presenting these results and observations is to better illustrate the nature of the science-practice gap and illuminate what often are the seldom heard or poorly understood perspectives of the practitioner community. The intended targets of my criticisms (and those of many people I interviewed) are the widespread perceptions and claims that formal science has been, is, or will be of practical value in the practice of environmentally oriented applied disciplines. However, it is *not* my intention to criticize or question the value of the scientists and their formal research programs. On the contrary, I happen to greatly value and appreciate this kind of science—most of my own career has been devoted to studying, performing, and teaching it. I am proud of the "good science" I believe my colleagues and I have accomplished, and I have no regrets about what we have done or how we have done it. In fact, unlike many of the practitioners I interviewed, as long as it did not directly compete with or take resources away from their work, I would be in favor of *increasing* our overall support for more basic research in general and academic ecology in particular.

I am also well aware of the fact that some scientists have been, and increasingly are, working very hard to create better models, strategies, databases, and tools to facilitate more and better applied restoration and conservation, and undoubtedly some of this work will eventually help accomplish these goals. Moreover, like many of my colleagues, I believe that the pursuit of "pure" knowledge is sufficient justification for research;

whether or not it ever intentionally or serendipitously leads to anything of practical value is a separate issue. Indeed, many scientists I know and respect are motivated more by a basic intellectual curiosity about and love of nature than an explicit desire to solve applied problems. These scientists would thus be the first to admit that their research may never have much, if any, practical value because that is not even their intention. For example, here is how one such research scientist explained his orientation: "Maybe some day I'll be able to help bridge the science-practice gap . . . but frankly, I don't want to alter what I'm doing to solve more narrow, immediate management needs. Basic science is my passion—it's what I'm trained in, what I'm good at, and what I get paid for. And I think it's important even if it has no immediate practical value or relevance."

However, it has unfortunately become all too common to falsely justify past and future scientific research on the basis of its alleged practical relevance and value. Such false or inflated claims obviously widen the science-practice gap and diminish the legitimate value of this kind of science. As one person involved with assessing and regulating conservation programs put it, "I read the parts of those ivory tower science papers and listen to the parts of their talks in which they always say how valuable their work is or will be to the folks on the ground, and it's grating. I wish they would at least use language that was a lot more cautious and realistic . . . as it is now, the main thing I think it demonstrates is how irrelevant and disconnected those academics are from the conservation and management communities and the local culture."

As I often experienced myself, explicitly requiring or implicitly expecting scientists to discuss the practical value of their research can also put scientists who know better in an awkward and ethically difficult situation. As one scientist said, "I love doing my research—evolution has given us these big, curious brains that like to ask questions and try to solve obscure puzzles. But I *hate* writing those 'implications for managers' sections everyone always makes you write! I wish we could just skip all that and get back to our science and stop pretending that we're doing something that we all know we're not."

Yet ironically, perhaps in response to the widening gap between science and practice and the deteriorating ecological realities on the ground, scientific organizations, funding agencies, and academic institutions have increasingly asked scientists to demonstrate the practical value of their work and to participate in activities designed to inform and guide the work of practitioners. For instance, there are now many meetings and sections in larger, more general conferences explicitly devoted to fostering more communication and collaboration between scientists and practitioners. While these are obviously well-intentioned and important goals, in my experience some of these efforts have unfortunately had the opposite effect. Many people I interviewed were also disappointed with and frustrated by their experiences at various "bridge-the-gap" events. For example, one practitioner responsible for managing a large natural area put it this way: "It's a big deal for us to find the time and money to go to these meetings. We aren't trained to make PowerPoint presentations, analyze data, and give polished talks . . . it's really intimidating for us to get up there and speak in front of a large audience full of scientists. And most of these meetings are dominated by researchers anyway—researchers hosting panels, researchers leading breakout sessions, researchers giving talk after talk after talk. We look at all their data and listen to all their talks and think 'so what?' I admit a lot of it is cool, fascinating stuff, but it doesn't tell us anything about how to take care of the lands we are struggling to manage . . . In the past, when we complained, the organizers tried to help by sending out surveys asking us to list the kinds of questions and specific problems we'd like the scientists to help us with. We'd fill them out and send them back, but nobody ever touched them, probably because they didn't involve any cutting-edge science, so guess what? We all just stopped going to those meetings."

Often, for different yet somewhat parallel reasons, many of my scientific colleagues and I also found such meetings uninformative and unproductive. This was because on the applied side, that practitioner was unfortunately right: many resource managers are not used to public speaking; not surprisingly, their talks often *were* poorly prepared and presented. In addition, there usually wasn't much "real science" at these meetings, and thus we in turn didn't learn much that could help us tackle our own pressing technical problems or advance our overall research programs. Consequently, many of us stopped going to those kinds of bridging-the-gap meetings as well.

Perhaps in response to this trend, over time the research portion of the science-practice meetings appeared to grow ever more technical and abstract. Not surprisingly, once again this probably well-intentioned response often backfired.

For example, I once sat through an entire three-hour workshop at a Society for Ecological Restoration meeting in which a panel of distinguished scientists provided practical advice for improving on-the-ground restoration programs. Although several of these talks were delivered at a fast and furious pace, I tried to jot down the major arguments and suggestions of each speaker. By the end of the workshop, my four-page list of "essential things for practitioners to incorporate into their work" included the following phrases: spatially explicit landscape modeling; scale dependency; nonequilibrium paradigm shifts; nonoverlapping literatures; bio- and socioeconomic matrixes; the theory of island biogeography; minimum viable population densities; metapopulation dynamics; inbreeding depression; outbreeding depression; cultural sustainability; pre-Columbian influences; panarchy theory; biocomplexity; watersheds; genetic architecture; predisturbance baseline models; fractal fragmentation patterns; ecotonal phase transitions; fungal communities; long-distance hydrologic interference; inter-situ connectivity; philosophical integrity; and climate change.

Looking over my notes, I realized for the first time that the "management implications and recommendations" sections of our talks and papers largely consisted of ever more extensive and sophisticated lists of critical new things for these already beleaguered people to consider and address. Part of me wanted to stand up and shout: "And we wonder why more practitioners don't attend these sessions, read our literature, and consult us? I have a PhD in ecology and many years of academic and applied restoration experience, yet I wouldn't have a clue how to translate any of this stuff into a real-world management plan!"

And yet we tend to pat ourselves on the back for doing these kinds of "outreach" activities and cite the proliferation of new applied science journals and required "implications for practitioners" sections in our technical papers and talks as evidence of the practical value and collaborative nature of our research. But despite these and all our other good intentions, too often we have wound up confusing, overwhelming, intimidating, and even paralyzing many of the very same practitioners we have been trying to help. Thus, if we truly want to facilitate more and better ecological restoration, it is imperative that we come up with new and more effective ways to bridge this prevalent, persistent, and widening gap between science and practice.

Chapter 8

Bridging the Science-Practice Gap

In an ideal world, research ecologists would provide ideas, guidance, and rigorous data that benefit practitioners, and practitioners would put this science into practice, exchange insights with the scientists, and make their project sites and management plans available for them to develop and test their theories. Developing and strengthening this kind of positive, mutually beneficial relationship between the scientific and resource management communities has been a central goal of organizations such as the Society for Ecological Restoration (SER) ever since its inaugural meeting in 1988.

Indeed, many pioneering scientists firmly believed that the work of academically trained ecologists and on-the-ground practitioners was and should be closely related and interdependent. For example, the authors of the 1987 classic *Restoration Ecology: A Synthetic Approach to Ecological Research* wrote that "both the restorationist and the restoration ecologist seek to reconstruct the system—the one in order to conserve it, the other in order to test ideas or to demonstrate an understanding of it. Recognizing this, and taking advantage of it, might provide a solid basis for a closer, two-way relationship similar to the one that exists in medicine, where clinical work and basic research often proceed hand in hand."

Unfortunately, despite many well-intentioned efforts, the discipline and literature of restoration ecology and conservation biology have remained dominated by ecological studies conducted in applied settings rather than research programs and papers that inform and facilitate on-the-ground efforts. At the same time, because practitioners rarely perform and document the kind of "rigorous" investigations required by professional scientists, their work and literature have been largely ignored by academia.

Fortunately, however, there appears to be growing interest in both communities to recognize and address these problems. For instance, SER launched a web-based Global Restoration Network explicitly designed to "link research, projects, and practitioners in order to foster an innovative exchange of experience, vision, and expertise," and some recent academic books and papers have begun to acknowledge and analyze this gap between ecological scientists and practitioners in particular and basic and applied science in general.

In this spirit, prior to the 2009 SER World Conference on Ecological Restoration in Perth, Western Australia, several colleagues and I sent an online survey to all registrants for this meeting to learn more about their perceptions of, and ideas for improving, the science of restoration ecology, the practice of ecological restoration, and the relationship between these two disciplines. We were pleased that over 70 percent (381) of the people we contacted completed at least the multiple-choice portion of this survey, and many went on to provide detailed and insightful answers to our open-ended questions.

Analysis of this survey data confirmed our hypothesis that despite some recent progress, the science-practice gap remains a major barrier to more and better restoration. For example, when asked about their perception of the relationship between the science and the practice of restoration, only about one-quarter of the respondents believed they were "in general mutually beneficial and supportive." The science-practice gap was also the second and third most frequently cited factor limiting the science and practice of restoration, respectively ("insufficient funding" was first in both cases).

Are restoration ecologists ignoring the needs of practitioners, or are practitioners ignoring relevant science produced by restoration ecologists? Even though our survey population contained three times as many researchers as practitioners, respondents' comments clearly and consistently favored the practitioners. Virtually no one faulted practitioners for ignoring available science, but many criticized restoration ecologists for ignoring the pressing needs of practitioners, performing irrelevant research, or failing to effectively communicate their work to nonscientists. Some argued that ecologists still had far more to learn from practitioners than practitioners did from ecologists. I was also gratified to see that many of the researchers themselves claimed to be well aware of these problems and committed to addressing them.

This challenge of bridging the science-practice gap is by no means unique to restoration; in fact, people in fields ranging from agriculture to medicine have been struggling to connect science to the "real world" ever since disciplinary science emerged. Given the magnitude of today's environmental crises and the extent to which many decision makers continue to ignore the recommendations of scientists studying these problems, it is not surprising that bridging this gap has proven to be particularly difficult in disciplines that involve both the environment and a diverse assemblage of human stakeholders. For instance, conservation biology, which from the beginning similarly dedicated itself to an activist, problem-solving agenda, continues to struggle to close its own considerable science-practice gap.

What can we do to help develop and strengthen the mutually beneficial relationships between ecological scientists and practitioners that everyone claims to want? Some have argued that we must transform the mushy and messy discipline of ecology into a more unified "hard" science so that it may eventually become as powerful and useful to its practitioners as, say, physics and chemistry are to theirs. This argument is actually part of a much larger and infamous internal debate among scientists over the "right" way to perform, interpret, and apply the underlying science of academic ecology (it's often said that we circle the wagons and then start shooting at each other). For example, a prominent ecologist in 1987 began his review of the arguments within and among the numerous and often acrimonious factions that constituted the discipline at that time by writing: "Ecologists are in a period of retrenchment, soul searching, 'extraordinary introspection' . . . This follows on nearly three decades of the heady belief on the part of some ecologists . . . that communities are structured in an orderly predictable manner, and of others that information theory, systems analysis, and mathematical models would transform ecology into a 'hard' science." While much has changed in the years since this review was published, the soul-searching and the "extraordinary introspection" over the relevance and rigor of the myriad competing approaches to ecology remain.

As intractable as this debate may be, its difficulties are trivial compared with the challenge of transforming restoration ecology into a unified hard science. In addition to grappling with many of the same messy ecological problems of academic ecology, restoration scientists must contend with the even greater complexities created by the even messier world of humans. Moreover, deciding to "promote ecological restoration as a means of sustaining the diversity of life on Earth and reestablishing an ecologically healthy relationship between nature and culture" (SER's mission statement) in the first place requires a commitment to a set of personal, subjective values and collective political movements that lie outside the scope of science.

Indeed, many of the people who took our SER survey argued that one of the most important things we could do to help bridge the science-practice gap would be to create and support alternative, "softer" research paradigms and programs that more effectively inform and facilitate the work of practitioners and promote more open and honest exchanges between restoration scientists and practitioners. Some also stressed that restoration ecology is not rocket science—it's much harder. We don't deal with things like atoms and orbits that can be effectively isolated and modeled with machines and mathematical equations. There probably never will be a grand unified theory of restoration ecology to pursue or prestigious international prizes to win, so lose the "physics envy," get real, and get to work!

What these alternative research paradigms and programs might look like depends on whom you ask. Some believe that better integration between the bio-ecological and human-ecological aspects of ecological restoration would be achieved if we turned restoration ecology into a transdisciplinary "metadiscipline." Under this model, restoration ecology would effectively transcend the conventional "normal" sciences via a paradigm shift away from its current emphasis on reductionistic and mechanistic processes and toward a more holistic and organismic approach. Others have similarly argued that the dynamic, nonlinear nature of natural ecosystems demands more flexible, robust, and nonhierarchical alternative research models. For instance, some have championed "adaptive management" protocols, in which management actions are explicitly designed as scientific experiments so that their effectiveness can be continuously assessed and refined as necessary.

Yet while such reforms often sound good in theory, and may eventually lead to research that more robustly captures the complexity of the real world, at present there still are few professional incentives to orient even these nonconventional approaches toward directly facilitating more and better restoration and conservation and solving on-the-ground problems. Thus, even if one or more of these alternative models were one day widely adopted by research scientists, the net result could simply be the creation of new academic frameworks and jargon-filled practitioner recommendations that might not necessarily be more practically relevant or helpful than the ones they replaced.

I would therefore argue (as did some of our SER survey respondents) that the best way to support scientific research that is more practically relevant and directly applicable to real-world ecological problems is to provide the necessary incentives and rewards for doing so. For example, we could value and support on-the-ground accomplishments at least as much as we presently do conventional academic accomplishments, and we could create prestigious institutional positions and professional journals for exemplary ecological restoration practitioner-scientists and productive, real-world collaborations between academic and applied science. These reforms could also dramatically improve participation, dynamics, and outcomes at our various bridge-the-gap meetings and outreach activities. After all, academics and research scientists are by necessity highly skilled at following the jobs and money; in the end, despite all our inherent independence and contrariness, we'll do whatever it takes to please our employers and funding agencies.

Many who took our SER survey also emphasized the importance of securing more money and support for restoration as a whole. They pointed out that we still do not have nearly enough resources to perform the restoration science, practice, and outreach programs that desperately need to get done. The scientists and the practitioners in our survey appeared equally frustrated by continuing public ignorance of and lack of support for their work. Consequently, many argued that developing more broadly based political support for restoration is the single most important thing we can do to both advance our science and practice and help bridge the gap between these two disciplines.

Early in my professional research career, whenever I expressed frustration with the ongoing rapid ecological deterioration and the snail's pace of conservation, one of my supervisors would always say, "Your job is to do the absolute best science you can. If you do that, everything else will follow." Like most of my colleagues, at that time I was more than happy to stay out of what I considered the stinking cesspool of politics and focus exclusively on my pure, clean science. I also believed that my high-quality, objective data would ultimately help settle controversial and emotionally charged resource management issues. But after many years of repeatedly failing to help settle anything with my science, I realized that the conflicts I was trying to help resolve were in reality largely political disputes.

For example, when I first starting working at the Ka'upulehu Dry Forest Preserve, I thought that our rigorous documentation of the positive effects of excluding ungulates from that remnant forest would help resolve the larger debate over whether and to what extent these animals negatively affected Hawai'i's native biodiversity. Yet I soon discovered that the devastating ecological effects of ungulates throughout the Hawaiian Islands had in fact been well known for more than a century and a half and had been extensively documented in both technical and popular publications. This ongoing ecological devastation is also painfully obvious today to even a casual, semi-informed observer of native Hawaiian ecosystems.

I eventually realized that the real reason the ecological effects of ungulates in Hawai'i remain "controversial" is politics (particularly on the part of the hunting and ranching lobbies), and I had been naïve to think that my science or anyone else's, no matter how "definitive," could by itself ever resolve this conflict. A senior research scientist later told me, "We probably could have done what has made the biggest difference in conservation in Hawai'i just by following the advice of local, on-the-ground people who said fence here, weed there . . . we didn't need all this science to see the effects of the ungulates, hotels, and habitat alteration. We could stop all the science now and it wouldn't have much impact on conservation, because we're still stuck on the same political and cultural obstacles we were thirty years ago."

The general difficulty of trying to use science to resolve political controversies was eloquently summarized in an article published, ironically enough, in *American Scientist*:

> Scientific inquiry is inherently unsuitable for helping to resolve political disputes. Even when a disagreement seems to be amenable to technical analysis, the nature of science itself usually acts to inflame rather than quench the debate . . . Science seeks to come to grips with the richness and complexity of nature through numerous disciplinary approaches, each of which gives factual, yet always incomplete, views of reality . . . "More research" is often prescribed as the antidote, but new results quite often reveal previously unknown complexities, increasing the sense of uncertainty and highlighting the differences between competing perspectives.

What should our top conservation and research priorities be? Which alien species represent the biggest ecological threats? Is biological control a safe and effective management option? The more I observed and participated in the passionate debates surrounding such seemingly straightforward "scientific" questions, the more I came to see the extent to which their answers depended on one's politics and values. Moreover, once these debates moved beyond the scientific and practitioner communities and involved individuals from other disciplines, special interest groups, and the general public, they tended to center on ever more political, value-laden questions: Which exotic plants are safe to sell to the public? Is ecotourism an effective conservation tool? How much money should be spent trying to preserve and restore our remaining native biodiversity relative to society's many other pressing needs?

While members of the environmental community often argue that ecological controversies should be resolved by the "best science," what I think they may really want is not so much the science as the politics and the value systems of the scientists. Like environmentalists in general, scientists who study ecological systems tend to love and care about them much more than might a random sample of the public in general and politicians in particular (I know I do). These personal values in turn tend to predispose us toward "objectively" interpreting our technical data and observations as justifying more environmentally friendly policies and procedures and supporting proactive interventions such as restoration ecology.

I once gave a talk at a research university in which I tried to illustrate some of the ways in which academic scientific research (most of which was my own) failed to help, and in some cases actually hindered, a series of resource management conflicts and applied restoration projects. I concluded by suggesting that as trained scientists, we of all people should be willing to objectively evaluate the practical merit and overall relevance and effectiveness of our work. When I was finished, a scientist in the audience immediately jumped to his feet, shook his head disapprovingly, and said, "You sound like an ex-Catholic who has lost the faith!" Later, as I thought more about his comment, I realized that he was absolutely right—the assumption that science is centrally important to the resolution of environmental problems, and that it is, should be, or one day will be, the most efficient and appropriate approach to ecological restoration and conservation biology, *is* a matter of faith.

Many people, scientists and nonscientists alike, appear to have this faith and believe in the universal supremacy of the scientific method with a fervor that resembles religious fundamentalism. Some also subscribe to the theory that such science has been the driving force behind much of the "progress" humans have achieved over the past thousand years or so, and similarly assume that most of the past and present progress made by those outside the sciences is attributable to the "fact" that they have had the luxury of standing on the giant shoulders of the scientists who preceded them. They are thus absolutely convinced that ever more science is the best, or even the only, way we can solve our current problems and continue to progress into the future.

Yet many scholars have convincingly challenged these assumptions and conclusions. Some have also pointed out the dangers and inherent irony of "hard-nosed scientists" effectively treating science as a religion. For example, in his book about this subject, Mikael Stenmark observed that "some scientists seem to have an almost unlimited confidence in science—especially in their own discipline—and about what can be achieved in the name of science." However, he later pointed out that statements such as Francis Crick's claim that "we are nothing but packs of neutrons," Carl Sagan's statement that "the Cosmos is all that is or ever was or ever will be," and Richard Dawkins's "every living object's sole reason for living is that of being a machine for propagating DNA" are extrascientific or philosophical claims. That is, even though these statements were made by brilliant scientists, there is nothing "scientific" about them because they are based on nontestable, nonfalsifiable personal convictions.

Stenmark's concluding analysis of these issues is particularly relevant to the goals of building more political support for ecological restoration and closing the gap between its scientists and practitioners:

> The public has to be more suspicious about what is claimed in the name of science, and scientists themselves need to be less naïve about the impact of their own ideological beliefs or value commitments on their scientific theorizing. What is called science can be far from an objective and dispassionate attempt to figure out the truth entirely independent of theism and naturalism, or of political and moral convictions . . . It is the conflation of these elements that gives the false impression that science can be one's religion . . . the truly scientific mind must instead be conscious of the limitation of the scientific enterprise, and also allow forms of truth and knowledge which lie beyond the scope of the sciences.

Even a cursory review of the history and development of any branch of science will reveal how quickly dominant paradigms and implicitly accepted "truths" may change, and how much they can be affected by the politics and value systems of their times. Realizing that our present scientific knowledge and methodologies will almost certainly one day be radically modified or completely replaced by new ideas and approaches can help us avoid the arrogance and hubris that too often have characterized scientists and our science. A more humble, open-minded, and inclusive disposition could also help us better understand and appreciate the different perspectives and values of restoration practitioners and the general public. This in turn could help us collectively bridge the science-practice gap and develop stronger, more broadly based support for ecological restoration and conservation in general.

One approach that could help accomplish these goals is to put ourselves in the practitioners' shoes. Indeed, several practitioners whom I independently interviewed told me how much they would like to take scientists out in the field, show them what they do, and *physically* work together with them. One technician put it this way: "It would be so helpful if there was some kind of requirement that all field scientists go through a kind of 'management boot camp' in which they would have to experience our world. Most of them don't have a clue what it's like to camp out in the rain forest, build fences, snare pigs, dig holes in the lava for outplants . . ."

Although I never went through such a boot camp myself, I did find that spending time with practitioners in the field was one of the best ways to better understand and appreciate their work and build mutually beneficial and respectful relationships. In addition, such experiences were seldom a one-way street; my efforts to help practitioners better understand and appreciate the world of science were usually far more successful when I was on their turf rather than mine.

As an individual research scientist with a seemingly incurable habit of sticking my nose into the applied world, I discovered three more steps I could take to help me become a better restoration scientist and develop and maintain better relationships with practitioners.

First, I learned to scrupulously avoid making false or inflated claims about, or taking undue credit for, the practical relevance and value of my or any other scientist's past, present, or future research. The more I did this, the more I discovered that many practitioners are actually quite interested in and supportive of basic research as long as it is not falsely presented and justified as being critically important to their work or performed at the expense of urgently needed management actions.

Second, I advised some practitioners to spend far *less* time and energy on their often surprisingly intensive "scientific quantitative baseline data collection protocols." Typically, their agencies were already overflowing with what in reality were pseudoscientific, uninterpretable data that required enormous resources to obtain and process. Many of these practitioners privately admitted that they knew these data were largely useless, but for various reasons they felt obligated to go on collecting more (e.g., "Some hotshot scientist told us to do it this way fifteen years ago," "We're always being criticized for not being rigorous enough"). Consequently, they tended to greatly appreciate my support of their abandoning such protocols in favor of "quick and dirty" yet far more effective techniques, such as simply recording and tracking the effects of their management actions with a series of standardized photographs.

Third, I strived to treat practitioners with respect, minimize my demands on their time and resources, and "give back" whenever I could. Contrary to the stereotypical assumptions made by many scientists, these people often have quite hectic, stressful, and grossly undersupported jobs and thus are understandably irritated when we arrogantly expect them to drop everything for us or we assume that supporting our research is or should be one of their top priorities. I also tried to respect the biological and physical resources that many practitioners work so hard to preserve and restore. It is infuriating to most managers when we not only take up their time but also damage their resources without giving anything back in return. One practitioner I interviewed captured the comments of many of his peers: "Scientists need to do their work in a more ethical manner; they should have to take some kind of 'first, do no harm' oath before they're allowed to work in some areas. I've been involved with far too many research and monitoring projects in which we gridded up tens of thousands of acres of rain forest. Because the scientists who pressured us into doing these projects needed to implement their 'rigorous sampling designs,' we had to do things like establish plots in all the different ridges and gulches, east- and west-facing slopes, etc. They didn't seem to care or even notice that we ended up trampling some really sensitive areas to do that! So we bust up the forest, create all these new avenues for the weeds to come in, lose all that time we could have spent actually doing conservation, and what do we ever get back from these scientists for all our time and effort? Absolutely nothing! It would have helped a lot if they had found ways to give something back, and I don't mean their research publications! Getting us some funding, contributing some labor toward our projects, giving nontechnical talks, and writing simple, one-page pamphlets that help the public understand and support our work—those kinds of things would really help improve our overall relationships."

Because what makes people feel, believe, and do what they do is such a complex, personal, and often emotional subject, even seemingly like-minded individuals may have radically different perspectives on some of the most important ethical, philosophical, and practical issues surrounding the science and practice of ecological restoration. Consequently, people have many different reasons to do it, many different ways to do it, and many different ways to define and evaluate its success. Understanding, appreciating, and respecting these different rationales for and approaches to the science and practice of restoration ecology and conservation biology is yet another important step we can take to help bridge the science-practice gap and create a more inclusive environment for these disciplines.

The following are a few contrasting viewpoints that illustrate this exceptional diversity of perspectives. These quotes are excerpted from more extensive interviews I conducted with a broad swath of people working in Hawai'i's conservation, environmental education, practitioner, regulatory, and scientific research communities.

What motivates you to try to preserve and restore Hawai'i's native species and ecosystems?

"My interest is more cultural than ecological; Hawaiian species are the true natives—they were here before the first humans. After I started learning the Hawaiian language, working with native Hawaiian people, and learning what they thought of and how they used the plants, I just became engulfed in that whole perspective of having a cultural responsibility to love and care for the natural world."

"I don't want to have to pay five dollars for a gallon of desalinized water from the ocean, or have to start living with fire ants and snakes!"

"I'm always thinking about the way things were here before humans arrived—the wild and crazy paths evolution traveled in the absence of things that were present everywhere else—and I just want to preserve as much as I can of those things that make this place unique and special. So my motivation is purely aesthetic—preserving biodiversity just makes me happy; it's biophilia."

"As a graduate student, searching for hours in the rain and mud for some bloody little thing and seeing one, or two, or, more often, none of them led me to develop a special affinity for Hawai'i's most desperate cases. I just love all the underdogs, the species destined for extinction."

"I was raised a Catholic, but I've always seen God in nature and believed there was a higher power out there. I have a personal relationship with particular native species; being with them, and experiencing the natural balance and beauty of intact native ecosystems . . . it just gets me high."

"Whenever the park's resource management guys would come in to our gas station after work, they'd all be smiling, laughing, and having fun, and I thought, 'Hey, I want to do that!' I thought that getting to work outside, doing stuff like killing weeds and building fences, would be a lot more fun than fixing cars."

"In the future, people can look back and say, 'Hey, this plant was extinct in the wild, but people took the effort to save and restore it!'

That's all the motivation I need to keep going—I'd like to think if I did anything of value in this life it will have been to resuscitate some fossils."

"I believe we've mucked things up and have a moral obligation to try and fix them. I see us as the modern missionaries: we come into a place and play God, even though we obviously don't have all of God's knowledge—just enough to be dangerous!"

How would you describe your overall restoration philosophy and strategy?

"We have to first develop a deep understanding of a place's community structure, then patiently design a plan that takes into account fine-scale processes like moisture, substrate, nutrient and disturbance gradients, locally adapted gene complexes, biogeography, and evolution. We need to analyze, fine-tune, and integrate everything so that all of these critically important factors and processes fit together— that's what I'd consider real restoration that's worth doing!"

"The Hippocratic oath of ecological restoration? I don't buy it one bit—of course we're going to do harm! We've got to move fast and take big chances—we've got no choice. Ecological communities are not composed of any fixed, magical combination of species—they're not necessarily coevolved or 'balanced'; they're highly malleable, and they can survive and evolve with new players. So we've got to quit worrying about everything being so pure!"

"It's only logical that we try to save the best, most pristine places first. I know that the longer you wait, the more difficult and expensive restoration becomes, but I still wouldn't support doing it until all the relatively intact, high-quality areas are safely protected first."

"Don't give up on the "basket cases," because they're savable! Don't just go after the easy stuff; don't go for the triage model. Nothing is hopeless—shoot for the moon!"

"Most of society doesn't get and can't embrace the big conservation picture, and thus they can't understand the concept that everything is connected. So we should start by working in local neighborhoods . . . But we mostly do the exact opposite by focusing almost exclusively on the remote, pristine areas. Then we tell the public that these areas are so precious and fragile that nobody should ever go there except us— no wonder conservation isn't getting anywhere! To be effective and sustainable, it has to resonate on an emotional, cultural level, and it's got to have economic benefits to the local community."

"I know there are some places and projects that appear to have reversed the tide of degradation, but coming from the outside, it's clear to me those are at best a temporary blip in time. Like it or not, we need to be ruthless about our priorities and employ the triage model. Should we really be working in places where there are only five individuals of some species left?"

"People don't understand how variable things are over time and space—how so many times all you have is tiny populations in highly degraded remnant habitats, how so often there's not much else you can do due to politics and biology and limited resources . . . Yet outsiders come in and want to 'think systematically' and arrogantly criticize our postage-stamp projects and think they're going to accomplish some grandiose ecosystem-scale project . . . They just don't understand the critical importance of the partnerships behind those projects, or the time and energy it took to get even those little things off the ground."

"The great majority of people only care about the plants and birds. If we were really serious about saving Hawai'i's native biodiversity, we would all be working on arthropods. Their diversity is ten to thirty times greater than the next taxonomic group [flowering plants], and their threats are entirely different. We could preserve almost an entire assemblage of arthropods in a pretty small patch of forest relatively easily and cheaply, but no one ever even thinks of doing that."

"I've seen over and over again how critically important education is. Yet while everyone will say they think it's important, no one ever wants to do it! It's always the last little thing we tack on at the end of our talks and meetings and papers, the last thing to get funded, and the first thing to get cut."

While some of these different perspectives and methodologies are complementary, some are obviously contradictory and mutually exclusive. Not surprisingly, I found that the often clashing views of scientists and practitioners can be especially difficult to integrate within restoration programs. Indeed, some scientists apparently are unable, unwilling, and even uninterested when it comes to viewing the natural world through anything other than their formal scientific lenses. Such people also tend to derive great pleasure from the raw quantification of nature and believe that the quality and merit of any restoration project is a function of the quality of its data collection and analysis protocols. Consequently, they tend to have little understanding of, or respect for, those who see nature and our relationship with it very differently—for example, through a utilitarian, cultural, or spiritual lens.

On the other hand, some practitioners seem to care only about "saving" what remains of the natural world as quickly and efficiently as possible. They likewise tend to have little understanding of or respect for the world of formal science and its scientists, whom they perceive as too busy tagging and tracking some exotic flea to notice that if something isn't done fast, that flea's host bird and that bird's host tree will soon be extinct.

While the views of most people lie somewhere between these two extremes, it is nevertheless all too easy to consider one's own work as "real" restoration and criticize or ignore those who see and do things differently. Consequently, the inevitable arguments over the "right" restoration strategies and actions will and probably should continue, as they add yet another layer of passion and meaning to the whole enterprise. At the same time, however, perhaps the time has come for us to collectively acknowledge that despite all our differences, we actually have far more in common with one another than we do with the vast majority of people outside our disciplines. Perhaps we could also spend less of our precious time and energy on these kinds of divisive internal debates and more time and energy working together to build greater public awareness of and support for our work.

For instance, a few colleagues and I recently proposed the creation of a "Restoration Ecology Extension Service" modeled after the US Department of Agriculture's Cooperative Extension Service. We envision staffing this restoration ecology service with broadly trained, open-minded people who could both facilitate effective communication among the diverse members of the restoration community and provide the inclusive leadership and coalition-building skills that could ultimately result in greater political and financial support for all aspects and flavors of restoration.

While it is not always possible or even advisable to accommodate different philosophies and strategies within the same restoration program, most of us could get a lot better at the fine art of compromise. We could acknowledge to one another and ourselves that there are many different legitimate ways to define, justify, research, fund, implement, certify, and assess ecological restoration projects. We could also improve our ability and desire to look at nature and the human-nature interface through more than one lens.

Ken Wood, one of the plant collectors at the National Tropical Botanical Garden on Kaua'i, is a good example of someone who can seamlessly shift between scientific and nonscientific perspectives of nature. Ken is deeply interested in the world of formal science in general and plant ecology and taxonomy in particular. He is one of the best field botanists in Hawai'i and has discovered or codiscovered (usually with his colleague Steve Perlman, another superb field botanist and plant collector at the NTBG) dozens of new plant species. Yet Ken is also an intensely emotional and spiritual person who is often guided by metaphysical experiences that are far beyond the perceptions of most people. A good example of this kind of "Ken Wood experience" (which typically horrifies scientists and fascinates much of the public) is the day he made one of his most famous botanical discoveries. One afternoon he sat down with me on a bench outside the NTBG herbarium and told me the full story. "I discovered *Kanaloa kahoolawensis* on the vernal equinox in 1992 on an offshore sea stack on the south side of Kaho'olawe when that island was transitioning out of being used for bombing practice by the navy and eighty or so years of devastation from heavy goat and sheep browsing. We were working for The Nature Conservancy doing a biological survey of the island in case it ever got transferred back to the people of Hawai'i. [In 1993, after decades of lobbying by native Hawaiians and other interest groups, the US Congress finally authorized the conveyance of Kaho'olawe and its surrounding waters to the State of Hawai'i. This congressional act set the stage for an ongoing, multimillion-dollar ecological and cultural restoration program of the entire island.]

"We weren't finding much on the main island because of all that past devastation, but then I saw a sea stack through my binoculars that looked like it had an intact native shrubland. I'd never been over to that sea stack before because the logistics of getting there are very difficult, but we had ropes with us this time, so Steve [Perlman] and I wandered over. We tried to rappel over to it from another offshore islet, but it wasn't possible. But I really wanted to get over there; it was calling me.

"I looked out in the bay and there was a mother whale and calf out there playing. It was all about spring birth and life, and there was some really heavy energy. Even though the week before I had fallen down a waterfall and broken my middle finger and it was still all taped up, I just felt like I had to get to that sea stack, so I worked my way over and found a tree I could tie onto and figured out a way to get down by attaching some webbing to my rappelling rope.

"I worked my way around the sea stack until I reached a spot where I thought I might be able to scramble up and get to the place where the native shrubland was. I was climbing up and was a little scared, because it was about a forty-five-foot vertical climb and I was wondering how I was going to get back down, but then a real interesting thing happened. I'm not a religious person—I don't relate to organized religion, but suddenly a passage from Psalm 23 started playing over and over in my mind: 'Even though I walk through the valley of the shadow of death, I will fear no evil.' I made it to the shrubland, and it was full of native plants, including *Senna gaudichaudii*, which had never been found or recorded on Kaho'olawe before. And there was *Portulaca molokiniensis* everywhere, and *Panicum torridum* in flower with its really silvery leaves, and *Chamaesyce celastroides* var. *amplectens*, and only a few grassy weeds like *Cenchrus ciliaris*.

"Then I went up the hill and right there, immediately, I saw *Kanaloa*, and that's when that magic happens, when you see something that you know is different and time kind of stops and there's no voice in the background saying things like 'What am I going to have for dinner?' or 'I can't believe so and so said that to me.' There was no mind, the wind was blowing, I could hear the fluke of the whale flapping below me with the baby, and I could see my line dangling across the bay where I'd gone down. If you look at Kaho'olawe from the air, you see that it is shaped like a fetus, and where I rappelled is exactly where the umbilical cord would be . . .

"But back on the level of science, I was looking at the *Kanaloa* and I knew it was really interesting and unique—right away I was thinking this must be a new plant genus. There were two of them there, and there was seed, which we know now is a very rare thing. So I was able to collect seed and flowers that very first time.

"There was also a tricolporate pollen that people were calling the 'mysterious legume' because it dominates the microfossil layers that were laid down in the islands about 10,000 years ago, but no one could figure out what species it came from. But then I brought some *Kanaloa* pollen over to Fordham University, and we eventually figured out that this was the mysterious pollen, and the mystery was solved.

"Of all the places in the world to find a new plant genus, in such a battered and abused and ecologically devastated area . . . it just shows us that there is magic out there, and there's a lot more going on than we really know."

While I myself have had very little personal experience with, or even interest in, these kinds of mystical interactions with nature, I have come to better appreciate how important and meaningful they are for many others.

I did, however, have an experience in graduate school that forced me to view the world of science through a lens that I suspect few of my colleagues have ever considered.

On a typical summer "field day," I would rise at four o'clock in the morning, load up my little Toyota with too much gear and not enough food and water, and drive south out of Albuquerque in the still-cool darkness to the vast Sevilleta National Wildlife Refuge. When I reached the refuge's dirt road that led to my study site, a few pronghorn antelopes would often appear out of nowhere and race down the desert with me for a while until finally bolting across the road and out of sight. I was always happy to see them, as they were beautiful and graceful animals, but I inevitably grew impatient for them to go because they slowed me down and I was anxious to get to work.

My research site was dominated by creosote shrubs and various plants that grew directly beneath them and out in the more open, intershrub areas. One such plant, a desert mustard called *Lesquerella fendleri*, was the focus of my PhD dissertation. I discovered that both the *Lesquerella* seeds in the soil and the seedlings germinating and establishing themselves out of this seed bank were nonrandomly distributed across that area. In addition to these demographic spatial patterns, I found some nonrandom genetic patterns within and among the seed, seedling, and mature plant populations themselves.

I spent a substantial portion of my five years in graduate school trying to understand how and why these patterns formed and persisted over time. The fieldwork component of this research included marking, mapping, and measuring hundreds of *Lesquerella* plants and creosote shrubs in the field, collecting thousands of soil and plant tissue samples, and lugging these samples out of the desert to my car by backpack and plastic coolers.

One day, I took an old friend from out of town to my field site. We walked to the southern edge of my site, where the land dropped abruptly down into a broad, flat, picturesque valley with a sandy arroyo running through its middle. At the other end of this valley, the land rose steeply again to another flat plain.

"Have you ever poked around in those?" my friend asked, pointing to a series of cavelike openings on the far side of the valley.

"No," I answered, too embarrassed to admit that I'd been so obsessed with collecting my data all those years that I had never ventured past the point where we were standing.

We spent the next several hours exploring that valley, walking up and down its meandering arroyo and hunting for artifacts in what turned out to be some pretty fascinating caves. Near the end of the day, we climbed up the steep southern slope to see what we could see. When we reached the top, my friend couldn't take his eyes off the countless miles of breathtaking desert that stretched out below us to the south. But I turned around to look at my study site from this distant and novel perspective.

After a few minutes of admiring its raw beauty, I suddenly saw a ghost-like image of myself frantically scurrying around. I watched as this apparently possessed creature collected soil samples, tied flagging tape and forestry tags to every creosote shrub in sight, and numbered, labeled, measured, and recorded just about everything else he could reach. Then in my mind's eye I saw this character racing around the university, grinding up and freezing the plant tissue; performing daylong genetic analyses in the lab; sieving and spreading the field soil across countless rows of plastic flats in the greenhouse; and entering and analyzing the enormous amounts of data generated from all this work.

I stood there and watched this vivid image of myself with a mixture of awe and disbelief. Why on earth would any rational, ostensibly intelligent human being voluntarily choose to do all that? Come to think of it, why were so many of us choosing to do this? (There was nothing special about me and my research—most of my peers and professors were working at least as hard on their own insane projects.) What an odd thing to do with one's life, and what a strange way to try to learn about the natural world!

For the record, I believe in the end I did learn a tremendous amount about the natural world from doing all that "crazy work," and I remain proud of all the scientific publications that emerged from that research. However, every so often since that day, when I am observing someone else's obsessive research program or engulfed in my own, that same eerie, "what a strange thing this science stuff is" feeling will creep over me again, and I can more fully appreciate how bizarre this world must appear to the uninitiated and nonbelievers.

Yet ironically, I think this experience also made me a better scientist because it helped me more objectively evaluate the strengths and limitations of rigorous, reductionist science and more willing when necessary to consider other methodologies. After many years of observing and evaluating other people's work and experimenting on my own, I ultimately concluded that an alternative, less formal approach to sciences such as restoration ecology and conservation biology can often yield more practically valuable and timely results and more effectively bridge the science-practice gap within these disciplines.

Chapter 9

Intelligent Tinkering

What we do affects how we see the world and what we believe in, and how we see the world and what we believe in affect what we do. Because we all have such different perspectives of and experiences in nature, the science of restoration ecology and the practice of ecological restoration often serve as screens onto which we project our personal philosophies and aesthetics and values and metaphors.

For many of the people in the conservation community, restoration is like a never-ending battle in the war to save the "remains of a rainbow." For the practitioners desperately trying to restore severely degraded and endangered species and ecosystems, it can feel like being in the triage room of an underfunded and understaffed hospital in the wake of a never-ending catastrophe. For some of the bureaucrats, the massive amount of coordination, resources, technical expertise, and political commitment necessary to make restoration work at larger spatial and temporal scales may be more analogous to a complicated engineering project like landing on the moon or building an atomic bomb. For the applied scientists, restoration is often like trying to comprehend and reassemble a mysterious broken machine with several pieces missing.

Consequently, ecological restoration can be comprehensible or intractable, beautiful or ugly, and inspiring or depressing; what is appropriate and effective at one point in time and space may or may not be in another. Thus, we need a great diversity of metaphors and perspectives to perceive and practice restoration because one vision or approach does not encompass all. We also need a healthy diversity of basic and applied restoration scientists and practitioners (and economists, educators, philosophers, and so on) with different goals and values employing different methodologies and techniques. Nevertheless, I have consistently found that the most successful restoration programs typically design, implement, and refine their projects by utilizing a disciplined yet flexible and holistic approach that, as explained at the end of this book's introduction, I call "intelligent tinkering" in honor of the pioneering ecologist Aldo Leopold.

More than seventy years ago, Leopold maintained that ecology could and should be the "fusion point of science and the land community." Not surprisingly, he frequently railed against what he saw as the rigid and counterproductive disciplinary boundaries of his day and called for a "reversal of specialization" to counteract what he believed was our increasing tendency to learn "more and more about less and less." Moreover, he presciently argued that solving complex environmental problems necessarily requires (1) integrating knowledge from a broad array of disciplines; (2) incorporating intuition, ethics, and other "unscientific" modes of perception into our work; and (3) infusing our ecological science and knowledge with a sense of wonder and passion and a commitment to promote a more meaningful and responsible relationship between people and nature. Leopold also provided a firsthand demonstration of how to implement these ideas on the ground through his successful intelligent tinkering approach to the restoration of his own degraded farm, long before the actual science and practice of restoration ecology emerged.

In the modern world of applied restoration programs, intelligent tinkering similarly combines attributes of good science (e.g., objectivity, hypothesis testing, and rigor) with attributes of good practice (e.g., technical skill, local knowledge, and relentless passion). Also much like Leopold, rather than excluding or discounting the real-world complications that are so often critically important to the ultimate success or failure of restoration projects—for example, fine-scale ecological heterogeneity, local politics, and logistics of volunteer coordination—today's intelligent tinkerers explicitly incorporate such factors into the design and implementation of their experiments and management actions. While the ecological, bureaucratic, socioeconomic, and interpersonal dynamics of such projects are often unique, I have found that the people who design and lead them tend to have most or all of the following three traits in common:

1. They are open-minded, pragmatic, optimistic, charismatic, passionate, and persistent.
2. They begin cautiously, with small, tentative exploratory steps, but as their programs progress and their knowledge and experience increase, they are not afraid to implement bold, large-scale actions whenever they feel the potential benefits outweigh the risks.

3. They are at least as passionate about the human dimensions of their work as they are about the ecological dimensions. Consequently, they strive to incorporate substantial volunteer, educational, and outreach activities into their restoration programs.

The combination of these personal characteristics and an overarching intelligent tinkering approach can overcome even seemingly insurmountable ecological and human-generated barriers. To illustrate how all this can come together and work in the real world, here are some snapshots of the people and projects within three of the most effective and inspiring restoration programs in Hawai'i.

MONTANE RAIN FOREST RESTORATION

Hakalau Forest National Wildlife Refuge

Eastern Slope of Mauna Kea Volcano, Island of Hawai'i

The Hakalau Forest National Wildlife Refuge was created in 1985 under the authority of the Endangered Species Act. Hakalau (Hawaiian for "place of many perches") was the first refuge established in the United States to preserve native forest birds and their habitat. Today this refuge comprises almost 33,000 acres between 2,500 and 6,600 feet. Although the mid-elevation, more mesic region of this area provides the best abiotic conditions for the birds, by the time the refuge was established, more than 200 years' worth of damage caused by cattle grazing, logging, fires, invasive weeds, and feral pigs had largely converted a magnificent native rain forest into a vast degraded pasture dominated by monocultures of noxious alien grasses.

Jack Jeffrey, the refuge biologist, and his colleagues decided to try to reforest those montane pastures and reconnect them to the relatively intact lower-elevation rain forests. "We wanted to create higher-elevation, high-quality habitat for the native birds as quickly as possible," Jack told me (if global warming results in mosquitoes surviving at higher elevations, the birds will also need to keep moving upward to escape avian malaria). "But we knew that the birds wouldn't leave the existing intact forest and fly out over those open, treeless pastures." He was well aware that no one had ever attempted such an ambitious restoration project in Hawai'i, and no one had a clue how to do it or even whether it could be done.

Fortunately, Jack is not the kind of person to use our collective ecological ignorance and inexperience as an excuse for inaction. As I quickly discovered when I began my own research program at Hakalau, Jack is very supportive and appreciative of academic science in general, and of the considerable amount of formal outside research performed within his refuge in particular. However, when he couldn't find any published research to help solve the problems at Hakalau, he launched a series of informal experiments on his own. "For us," he explained, "it really has just been a lot of seat-of-the-pants experimentation and learning as we go."

When the thick glaciers high up on Mauna Kea began receding at the end of the last ice age, the resulting meltwater carved narrow gulches out of the underlying lava substrate as it flowed down the massive volcano toward the sea. Portions of these gulches apparently proved to be steep and deep enough to protect the vegetation within them from the human-induced forces that eventually destroyed the surrounding forest. Thus, today some high-elevation sections of Hakalau serendipitously contain elongated slivers of relatively intact native forest rising out of the surrounding ecological wasteland.

Inspired by what they saw in those tree-lined gulches, Jack and his colleagues decided to try to establish a new network of forested corridors. They discovered that, just as birds on the mainland often fly along the rivers that flow through highly developed or degraded landscapes, many of the native forest birds were similarly utilizing Hakalau's old-growth tree corridors to traverse the refuge's treeless upper-elevation sections. Why not capitalize on this phenomenon by planting long corridors of native, fast-growing Acacia koa trees at regular intervals across the entire refuge?

"We started planting the koa corridors in 1989," Jack recalled, "after we got the first subunits of the refuge fenced and the ungulates out. We also thought that maybe if we planted them close enough together, the forest might 'naturally' begin to fill in between the corridors and eventually shade out the alien grasses. Of course, we were talking about the 500-year plan here, not the 50- or even 100-year one!"

Once they got going, Jack, his colleagues, and an ever-growing band of volunteers never looked back. They also repeatedly solved their own problems by relying on observations and informal experiments to develop and refine their methodologies. For instance, Jack explained that when they first started, the standard koa-planting protocol worked, but "if we followed it, and planted the koas in big pots like we now do for some of the endangered species, we could plant maybe three trees a day. Well, we started experimenting to see if we could find a better way. We eventually found that with a sod-scraping three-pronged rake mounted on the blade of a bulldozer, we could plant 2,000 trees a day [grown in long, narrow dibble tubes instead of regular pots] using unskilled volunteers. This also gave us a path to walk on and follow, which is essential when you've got people stumbling around through all that tall grass out there. The scraping also kept the grass away for a year or two, which proved to be enough to get the trees established. So we just followed the path of least resistance up and down the pastures, which was generally along the contours, and the volunteers didn't even have to think, just follow the line and plant!"

In addition to transplanting thousands of these hardy koa trees, they began to grow and outplant other native species as time, labor, seed availability, and greenhouse space allowed. To direct these efforts, in 1996 the refuge hired Baron Horiuchi as its horticulturalist.

"When I first started," Baron told me, "I tried to look for basic information in books and papers, but a lot of these species had never been propagated. So I decided to just look at what nature was doing and try to imitate that, and for the most part that worked out pretty well. But then I started working with the endangered plants, most of which rarely if ever regenerate in the field. So I would go out into the forest and find and collect as many ripe fruits as I could before the rats got them. Then I basically just did lots of trial-and-error experiments using different organic substrates, potting mixes, and field soils, and waited to see where germination was best. It was effective, but it took a lot of time because the results were often completely different for each species."

By 1997, Baron was growing about 20,000 koa and 5,000 non-koa seedlings each year. One of the biggest challenges at this stage was to figure out how to successfully establish many of these relatively slow growing, less hardy non-koa plants in the field.

"When we first started," Baron said, laughing, "we tried just about everything! Two of the biggest problems were the frosts and the droughts. During El Niño years, we might not get a drop of rain for three months, then over thirty inches of rain in one weekend, and then another three months with nothing."

"We thought most of those species were going to do fine out there in the pastures," Jack added, "but in the end nothing survived in the open pastures—zero! But we had 80 to 90 percent survival when we planted within our koa corridors. I can't say it was the reduced competition from the grasses because the grass cover wasn't much different beneath the koas from what it was adjacent to the corridors. I think it was probably just the exposure and the tremendous frosts. In any event, we eventually figured out that as long as we had at least 25 percent overhead cover of koa canopy, the survival of our non-koa species was good, so from that point on we just planted everything under the corridors."

Thus, the koa trees turned out to be, as Jack calls them, the "forest engineer." Without the protection provided by their canopies, it was virtually impossible to establish the other native plants in all those exposed pastures, but by utilizing their ever-expanding network of koa corridors, large-scale rain forest restoration suddenly seemed very possible. In about a dozen years, they managed to establish these corridors (spaced 100–200 meters apart, depending on the topography) all the way across the pastures and up to their fence at the top of the refuge. By the summer of 2004 they had planted more than 350,000 koa trees and tens of thousands of non-koa plants.

When I asked the refuge staff members how they managed to implement such a large and successful restoration program, the first thing everyone said was, "The volunteers." They explained that with their tiny staff, modest operating budget, and myriad other responsibilities, there was simply no way they could have attempted such an ambitious restoration project on their own. Each staff member also stressed that in addition to labor, the refuge's volunteer program generated a tremendous amount of public education and outreach, which in turn supported their restoration program in many direct and indirect ways.

"Sometimes I really don't understand how strong human spirits are," Baron added. "Our volunteers come here from all over the world, they work eight to ten hours per day, in difficult terrain under difficult conditions, and they give so much of themselves to the land. It's hard letting some of them go because we become so close working together . . . I have a hard time putting into words how I feel about it. I tell them at Hakalau there are only good feelings and spirits because all these good people come up and help and leave all this good energy behind."

"We get so much out of them," Jack agreed. "But it also takes a lot of leadership and love for what you're doing to inspire the volunteers to do this kind of work. Some of them are already very dedicated before they get here, and thus you can use and abuse them and they keep coming back for more year after year. But we also get lots of people who are not used to being out in the field—for example, office workers from Honolulu, who pay their own way over and have to get up at four o'clock in the morning on O'ahu to make it here on time. We learned not to work them until six in the evening in the rain and cold on that first day! Now we might take them birding at first, give them a T-shirt . . . once we started doing those kinds of things, our volunteer applications skyrocketed.

"Now we are getting more and more groups coming here from the mainland," Jack continued. "They include everyone from hard-core wilderness freaks to CEOs of major corporations. Many of these people would be staying in five-star resorts if they were vacationing in Hawai'i on their own, yet they come here and say, 'You have my life for a week—what challenge are you going to give me?' They end up working their asses off for us, yet when they leave they always tell us how wonderful it was, and we get lots of pats on the back.

"In the beginning," Jack concluded, "I was very pessimistic because I was going to areas where the birds were there in the not too distant past, and now they were effectively gone. Seeing the birds dropping out right in front of me made me very much a pessimist. But then I got back into the research and management side of things, and I started to see that yes, we can do things that make a big difference, and we can turn this whole thing around. The success we're having here at Hakalau is very encouraging— we've got over 5,000 acres now in corridors, and most of the fenced area is finally ungulate free. And, of course, the best part is that the birds are coming back—that's the real proof that we're doing something right."

DRY FOREST RESTORATION

'Ulupalakua Ranch

Auwahi, East Maui

In the botanical literature, Auwahi is a centrally located 5,400-acre subsection of the southwestern rift of the Haleakalā Volcano on East Maui at an elevation of 3,000 to 5,000 feet. The famous botanist Joseph Rock identified the remnant dry forests of Auwahi and North Kona on the Big Island as the two botanically richest regions in the Territory of Hawai'i. The tragic story of the subsequent devastation of Auwahi's native flora largely mirrors the fate of the dry forest ecosystems Rock knew and loved on the Big Island: cattle ranching, imported grasses, fire, and an ever-expanding tide of noxious alien plants and animals. Another species in the same genus as fountain grass, kikuyu (*Pennisetum clandestinum*), also eventually dominated much of the formerly forested regions of Auwahi after ranchers deliberately introduced it in the late 1940s.

"I started working up in Auwahi in 1981," Lloyd Loope, the director of Haleakalā National Park's Biological Resources Division (its research arm) told me, "and I needed to find somebody who really knew their plants. When I asked around, everyone told me I should hire Art Medeiros.

"Back in 1967, The Nature Conservancy made the first attempt at conservation in Auwahi by constructing an ungulate exclosure, but then the kikuyu grass took off in there and it was just impossible to accomplish any-thing. They eventually let the cattle back in, and most of the conservation community gave up all hope of ever restoring that area. But then, fortu-nately for us, the sugarcane aphid got into Auwahi and started killing some of the grass, and then the Native Hawaiian Plant Society built a series of small exclosures to protect patches of some remnant dry forest trees. In the early 1990s, we saw some reproduction in those exclosures, and Art said, 'If we can keep out the weeds, maybe it's time to think about restoring this place.'"

"I remember the day when a coworker and I were having lunch at Auwahi, taking a break from some hard physical work and admiring the beautiful ocean," Art Medeiros told me. "He pointed down to the ten-acre exclosure we were trying to restore and said, 'Wow . . . what if this works?' I thought about that for a minute, then said, 'What if it works and no one cares?'

"So I started recruiting volunteers. No one came at first, then they started coming—all by word of mouth—and now there are so many we have to turn a lot of them away. Every time, I bake them cookies, and every time, I share a piece of my soul with them, because this restoration is spir-itual stuff, and you should get something out of it as well as give, so I work them hard, and I tell them they're doing something that's going to change the world: 'That plant that you are planting, it's going to live longer than you are, and it's going to have more babies than you are; it probably will change the world more than you are—its DNA is not going to go extinct—you're starting a DNA lineage on this land.'"

Although Art still works for Loope's Biological Resources Division, he is one of the very few government scientists I know who consistently prior-itizes and incorporates on-the-ground conservation into his professional work.

"I love science," he told me. "It's a great tool to add to our knowledge. And I think I know how to be a good scientist. But I'm not interested in having a big plump résumé or advancing professionally through the sys-tem. I'm much more interested in conservation and on-the-ground ac-complishments, connecting with kids, mentoring young people, building community . . . that's the stuff that matters to me!"

The combination of Art's drive and the strong support of the local land-owner ('Ulupalakua Ranch) led to a multiagency cooperative effort to re-store a small but botanically rich section of Auwahi, which began with in-stallation of a ten-acre exclosure. Despite several extreme ecological, financial, and logistic challenges, by the late 1990s the group had largely controlled most of the invasive weeds within that exclosure, constructed a greenhouse, and successfully propagated thousands of native plants.

For Art, "doing restoration right" includes both carefully navigating his way through the endless labyrinth of ecological and philosophical co-nundrums typically associated with this work and devoting a substantial amount of thought and energy to its cultural and spiritual dimensions. Thus, when his group was finally ready for its first major outplanting, Art invited a famous Maui *kumu* (teacher) to bless the plants and welcome them back to the *ahupua'a* of Auwahi.

After a round of prayers, the *kumu* mixed *'awa* root (*Piper methysticum*; the Polynesians brought this mildly narcotic shrub in the Pepper family with them to Hawai'i because of its cultural importance) with water in a heavy wooden bowl and then sprinkled the *'awa* water around the green-house plants. Thus blessed, the seedlings were loaded onto trucks, which carried them back to Auwahi. When the group later reconvened inside the exclosure, another Hawaiian man blew the *pū'olē'olē* (conch shell) once for each of the four cardinal directions. Art later wrote that its

> loud brave cry filled the emptied forest, echoing off its rocky ridges. I found myself wondering how long it had been since the *pū'olē'olē* had sounded at Auwahi. One hundred years? Two hundred? Three hundred? More? Maybe that was the reason the dryland forest at Auwahi had fallen on such hard times! . . .
>
> As I watched the thin milky *'awa* water being poured from the coconut cup into the planting hole, I felt I was watching the *ola* (life) being poured back into the land. I had always thought the *ola* was in the plants, but now I felt the *ola* was in the land itself, await-ing the arrival of the seedlings.

"Look," Art once told me emphatically. "I'm not some New Age crystal gazer; that stuff dishonors truth. But when something is real, it's real."

One Auwahi volunteer described how Art "works his routine magic": "He's so good at telling stories. Each time it's a different story, which goes in different directions, but it's always some story about Auwahi, and you're in the place, and you try to envision it as it was, and how we got to where we are today, and what we're going to do today to change things, and every-body just gets totally inspired."

"They come because they want to do something real," Art explained. "Most have highly tuned bullshit detectors—they look me in the eyes, weighing it all. Some of them are quite wealthy, and they want to know how this restoration thing compares to their Lexus and other material toys.

They'll start out pretty cynical, wondering how long they have to work and when they get to leave. You can't convert them all, but by the end of the day, for most of them, their cynicism is broken, and they end up sounding like kids in their optimism. They'll say things like 'Hey, we can't leave yet; we haven't finished weeding this section' and 'If we could do this every day, we could do hundreds of acres!' and 'Now we don't have to go to church tomorrow!'"

During my first visit to Auwahi, when the group was just getting the restoration program up and running, Art pointed out several instances of "natural" native seedling recruitment within the exclosure and explained that this probably hadn't happened on East Maui for at least some of those species in fifty to one hundred years. When I returned to Auwahi a few years later, as we approached that same exclosure I was immediately struck by a large square of vibrant green that jumped out of the surrounding drab landscape like a picture in a child's pop-up book. Art looked at me and beamed. "I deliberately made it square so that people wouldn't mistake it for some kind of natural feature. You can actually spot it now when you're flying over this part of the island in an airplane. That's one measure of success I thought I'd never live to see."

He explained that shortly after my previous visit, they had hand-sown over a million seeds of 'a'ali'i (*Dodonaea viscosa*, a relatively common indigenous shrub in the Soapberry family) to "occupy the beachhead" created by removal of the kikuyu grass, help suppress the newly emerging noxious weeds, and create favorable microhabitats for the reestablishment and spread of other native plants and arthropods. They had also outplanted thousands of individual seedlings and saplings from many other native species. Now there were several places within this exclosure dominated by such a thick mass of young native plants that it was difficult for even Art to tell which had been seeded, which had been transplanted, and which had come up on their own. When I pointed to a clump of mature, fruiting 'a'ali'i shrubs and inquired about their origins, Art just shrugged. "I don't know," he said proudly. "Many of our babies have been having their own babies for quite some time now."

When I asked him how he figures out what the right restoration mix of native species should be, he just threw up his hands and shrugged. "It's unknowable. But the truth is, I don't even care anymore about what the 'correct' proportion of each different species might be—I'm fighting a war! The purpose of my dry forest work here is to reestablish regimes of competition between native species; I'm willing to trust the balance that emerges from that process. Even if I wanted to, which I don't, there's no way I could micromanage and control that balance anyway."

Art grew increasingly animated as we wound our way down deeper into the exclosure. Walking through the "forest" with him was like walking with an expert curator through a natural history museum in which all the specimens had come to life, broken out of their display cases, and started interacting with one another. "Look at this vignette!" he'd say, pointing, when he came across a scene or specimen that he found especially interesting.

"That's the remains of a long-dead *Partulina* snail," he said as we made our way through the lush vegetation, pointing to a half-inch-long bleached shell lying in the leaf litter beneath a large native tree. "Its presence here indicates that at one time this must have been a much wetter and more densely forested site than it is today.

"Listen to those *'amakihi!*" Art smiled as he pointed to a group of native honeycreepers in the branches above us. "I've been hearing a lot more of them since we started the restoration—they're even nesting in here now.

"Take a whiff of these," he said, handing me a bunch of colorful flowers with a strong, pleasant citrus scent. "And that sweet, honeylike odor that's been following us around? That's the *maile* [*Alyxia oliviformis*, an endemic vine in the Dogbane family]. Probably no one has experienced this combination of colors and smells for hundreds of years, or maybe ever, given all the changes that have occurred since these plants and people coexisted. That's another goal of mine—to restore the colors and perfumes that have been lost from this landscape."

Art explained that part of his vision included one day being able to responsibly extract various materials from the forests of Auwahi, make things out of them, and maybe even sell some of the resulting products to help support the restoration program. "Sure, I want to have preserves where trees rot and nothing is taken. But I'd also like to have other places where we use the forest. You know, the ocean was the Hawaiians' refrigerator, but these dry forests were their toolboxes and medicine chests. Take the *'a'ali'i*, for example. They used its beautiful, strong wood for house construction, weapons, and farming and fishing tools. They used its leaves for medicines. And they made leis and a bright red dye out of its fruit capsules.

"Remember this," he added, spreading his hands wide to encompass the whole area, "because you won't recognize it the next time you're here. We've basically weed-proofed the first exclosure, and now we're on to the second. Only this time, by applying what we learned in Auwahi One, I'm confident we can do Auwahi Two for about one-fifth the cost, and about five times faster. I know what to do now and how to do it. I've got a great group of dedicated volunteers. And being able to finally hire Erica has been the icing on the cake."

Having experienced the fast, furious, and far-ranging experience of being with Art in the field before, I had sought out Erica von Allmen, Art's sole paid staff member, prior to our Auwahi tour. But when I asked her for a slow, comprehensive overview of their restoration program, she just laughed.

"We used to be a lot more structured in the beginning—we had our little trials and more formal scientific experiments—sun versus shade, seedlings inoculated with mycorrhizae versus no inoculation, and so on—but it eventually just became a matter of getting as many plants in the ground as possible. Except for the really rare stuff, we don't even bother to tag and track them anymore. I'm sure it'd be valuable and fun to do more monitoring and experimentation, but we're just too busy planting, and all that other stuff takes so much time and effort and money. We don't have any formal management plans or visioning documents for Auwahi, either—Art just constantly assesses the situation and makes his decisions accordingly. He knows what he is doing, and everyone respects that."

I asked Erica if she had learned any concrete lessons from her work on the Auwahi project.

"Well," she said finally, after thinking about it for a while, "I do see different levels of complexity now that I didn't see before. There are just so many interacting factors that are virtually impossible to track and tease apart and really understand. For example, are we imposing an artificial selection regime by always selecting the plants from the fastest-germinating seeds in the greenhouse? What's happening to the genetic diversity of our rare species and our dry forest flora as a whole? Will some of the species that appear to be doing so well now eventually not make it, due to the rats, or the loss of their original pollinators, or something else that we haven't thought of or can't foresee? But we're always stretched to the breaking point just keeping the whole operation afloat. And even if we ever had the luxury of focusing on those kinds of things, I'm not sure what we could really do about them anyway."

When I asked Art what lessons he had learned at Auwahi and elsewhere over his long and illustrious career, he immediately said that solving the ecological problems is actually less difficult than solving the human problem.

"What really discourages me is the way the people who are supposed to be doing something about all this clash with each other, and that so many of them have given up hope. That loss of hope is really our biggest enemy. Successful conservation requires a strong human presence—a champion. When we reach the tipping point of enough human hearts getting involved, really good things can happen."

MESIC FOREST RESTORATION

Limahuli Garden and Preserve, National Tropical Botanical Garden

North Shore, Kaua'i

Given Limahuli Valley's sheltering mountains, perennially flowing stream, frequent rains, fertile soils, and accessible, productive marine ecosystem, no one was surprised when recent archaeological evidence corroborated the local belief that this area had been home to one of the earliest settlements in all of Hawai'i. Today, the Limahuli Garden and Preserve consists of approximately 1,000 acres with three distinct subdivisions: the lowland montane rain forest of the Upper Preserve, the mesophytic and lowland rain forests of the Lower Preserve, and the more formal seventeen-acre, publicly accessible botanical garden.

Juliet Rice Wichman donated the first thirteen acres of Limahuli to the National Tropical Botanical Garden (NTBG) in 1976, and her grandson Charles "Chipper" Wichman gifted the remaining acreage to the NTBG in 1994.

"Even though I grew up here," Chipper told me, "like almost everyone else, I really wasn't aware of what was going on out there in the field ecologically—I just knew it was all green and beautiful. But later, when I was searching for some direction to my life as a teenager, I spent half a year working at Limahuli, and I learned more about my grandmother's vision to turn it into a botanical garden. Although she wasn't especially interested in native plants, she was into Hawaiian concepts of land preservation and stewardship. Like the Hawaiians, she believed that we are a part of the ecosystem, and that our role is to help maintain *pono* [balance]."

Chipper eventually went back to school and graduated from the University of Hawai'i at Mānoa in 1983 with a degree in tropical horticulture. During this time, he also became increasingly interested in Hawaiian history and the native culture and language, especially after he met his future wife, Haleakahauoli, who was enrolled in the university's Hawaiian studies program.

"There was so much I wanted to do in the Lower Preserve," Chipper recalled, "but at the same time, we were also mandated to become financially self-sufficient, so I had to start by focusing on the visitor pro-gram to generate enough money to keep the garden afloat. [This program took off in 1997 after Limahuli received the American Horticultural Soci-ety's Natural Botanical Garden Award.] After that was up and running well, I got that $15,000 grant from the Forest Service for you guys to do that research in the Lower Preserve, so I held back to see how your experi-ments would turn out."

Chipper was referring to an experiment that a colleague and I had launched several years earlier. I had also managed to shake loose some in-ternal USDA Forest Service money for Limahuli to help defray the con-siderable in-kind costs associated with our research (chain saw work and brush removal, herbicide application, seed collection and propagation, nursery expansion and improvements, data collection, etc.). Together, we explicitly designed that experiment to produce what I hoped would be un-ambiguous, practically valuable guidance for Chipper and his staff as they scaled up their restoration efforts in the Lower Preserve.

While that research had gone well, once again our results had turned out to be frustratingly complex and inconsistent. In a nutshell, we found that different native and alien species sometimes responded differently to different restoration treatment combinations.

Years later, I asked Dave Bender, the Limahuli staff member who had done most of the fieldwork for that experiment, whether he had been able to extract anything from our research that was practically relevant to their larger restoration program. "Not that I can think of," he said. "In doing that kind of work, on that kind of scale, you mostly have to deal with the logis-tic complexities of getting people and equipment and plants in there. There just isn't much time or room for thinking about or doing much with those kinds of complex ecological interactions."

When I asked Matthew Notch, Limahuli's restoration project manager and volunteer coordinator, to reflect on his experiences in restoring the Lower Preserve, I got a similar response. "Scientific knowledge is obvi-ously important, and maybe down the road some day, when our program is further developed, we'll be able to utilize more of it, but right now it's re-ally just practical experience. We do very little monitoring or data collec-tion, unless we have to provide some formal documentation to a granting agency, because we're always just scrambling to get the work done.

"We really just learn by doing. We try plausible things, watch what hap-pens, and adjust our practices accordingly. Sure, we've found some gen-eral patterns, like some natives do better with full sun, and having some canopy shade can sometimes slow the understory weeds down and maybe let us get by with less watering. But what we actually do is pretty site specific. For example, in some areas where there are few or no natives, it's just easier for us to come in and clear out everything. We found that if we open it up, expose some bare soil, and stay on top of things with our follow-up weeding, we can get the naturally recruiting, pioneering natives to eventually dominate the understory and create good habitat for us to come in later and outplant the things that would never establish on their own. But we can only get away with this in places like where your research plots were, where the soil is fertile and there's an extensive native seed bank. Another quarter mile up the valley, it's a totally different story, so we had to figure out a totally different plan of attack up there."

As the restoration staff and volunteers continued to successfully scale up their mesic forest restoration work, they also began another ambitious project to restore the Limahuli Stream, one of the very few remaining Hawaiian streams with both high water quality and all five species of native freshwater fish. Adult ʻoʻopu, as the Hawaiians call them, live in freshwater, but their fertilized eggs wash downstream and the newly hatched larvae spend their first several months in the ocean. To complete their life cycle, these young fish must find and ascend a suitable freshwater stream. However, the restoration team discovered that these fish couldn't get past a section of the Limahuli Stream that was choked by a dense thicket of *hau* (*Hibiscus tiliaceus*). This small tree in the Mallow family is widely distributed in coastal and riparian areas throughout the tropics and subtropics; whether it reached the Hawaiian islands on its own or was brought over by the Polynesians remains unresolved. Apparently, the special suction cups of the ʻoʻopu were foiled by a thick layer of silt that the slow-moving water had deposited over the rocks in this section of the streambed.

"There was a half-acre jungle of impenetrable *hau* blocking the stream," Matt explained, "and it was a silty, mosquito-infested mess. At first we talked about air-dropping some heavy machinery in to clear it out, but then we were told we couldn't do that because of all the archaeology sites around there, and because that machinery probably would have silted up and damaged the streambed even more. So in the end we just got a big crew of volunteers together and used brute force. We started with handsaws and kept knocking stuff down and pulling it out until the water finally started trickling through again. Then we got some big floods that washed away the stumps and pushed out the silt, and the stream really starting flowing hard again. Now it's back in its old channel, and within less than a year we found that all five ʻoʻopu species were making it back up to the waterfall."

"It's amazing what you can accomplish when you have the will and the people to do what it takes to get the job done," Chipper said when I asked him to reflect on his experiences at Limahuli. "You can do more than you would ever think is possible at first. The hardest part was just getting that vision off the ground, but once we did, things started rolling. Now the VIPs go out there and are blown away by what we've done, and it's a relatively easy sell to get more money and keep the programs going. There's nothing like success to create more success!

"But also, there is often a lot of tension among the different components of all the things we are trying to do. When you have a very limited pool of money and resources to draw from, should you launch a scientific research project or break out the chain saws and Weed Eaters? Looking back, I'm glad we started with the science. But as we started getting more grants, and more money from our visitor program, I didn't feel a need to keep doing the science, because restoration is really more of an art than a science.

"For example, when we started restoring some ancient rock work in the Lower Preserve, we worked out an arrangement in which the native Hawaiian stonemasons and the archaeologists would have an equal voice, because as artisans the stonemasons could actually see things in the alignments of the stones and how they were set that the highly trained and educated archaeologists couldn't. Similarly, as the ecological restoration program proceeded, I realized that we needed to bring the practitioners into the visioning and planning processes, because so much of it is unpredictable, and we need to be able to adapt to what's happening on the ground. I had some pretty specific, detailed restoration ideas and methodologies in my original master plan, and we've since written many proposals in which we claim we're going to do things in a certain way. But one of the real challenges for me has been that when we started doing the actual work, the guys on the ground would often find that it couldn't really be done that way, or another way would be much better."

When I asked the Limahuli staff members what they felt their most important accomplishments had been so far, nearly everybody mentioned their education and outreach programs in general and their work with children in particular. "Almost every group that comes here gets engaged in a meaningful way," Kawika Goodale, Limahuli's assistant director, told me. "Different groups get hooked by different things—the majesty of the place, the medicinal value of the plants, the conservation mission, the cultural connections . . . But we've found it's best if we can get them working—touching and feeling things and getting dirty—and we always try to end with some planting. When they come back years later and see the effects of what they did, and especially how the plants they planted have grown ... there's just tremendous power in that experience; it's like a freight train!"

One day near the end of my time at the Kaʻupulehu Dry Forest Preserve, I stood on the edge of the highway and looked at the ecological devastation and ongoing degradation all around me. Despite all our accomplishments here — countless scientific talks and publications, on-the-ground alien species control and native outplantings, on-site tours and outreach efforts — no restoration had even been attempted on any of the adjacent lands. Moreover, the vast majority of people in the surrounding communities remained uninformed about the plight of Hawaiʻi's native dry forests and disengaged with our efforts to preserve and restore them.

Even if some group were to magically appear with the necessary time, interest, and means to attempt a larger-scale restoration project at Kaʻupulehu, what could we tell them now that we couldn't have told them when I first started working here? I thought about the concluding paragraph of our final report, summarizing what we had learned from our most recent round of research, which had been funded by a four-year, $270,000 US Department of Agriculture grant:

> We conclude that successful native dry forest restoration in fountain grass invaded sites will in general require grass removal, shading, and introduction of targeted native species. Because native plant responses to restoration treatments are often highly species specific, the most effective treatment combinations for a given individual species will most likely depend on its specific morpho-physio-phenological characteristics.

Once again, I was proud of our research, the specific knowledge we gained, the more general academic contributions we made, and the indirect yet substantial ways in which our work contributed to the North Kona Dryland Forest Working Group's restoration and outreach programs. Yet as valuable as all that science had been, I was also painfully aware that we had not discovered much that was of direct practical value to the restoration of this and other degraded ecosystems. I knew it was unfair to criticize our research for this, because the formal scientific methodology we had rightly employed was designed to yield "good science" rather than practically relevant and valuable results.

Nevertheless, I tried to visualize large numbers of dedicated people out on this landscape, earnestly striving to improve and expand upon the restoration accomplishments of our working group. But who would lead them? How could such a large group come to consensus on the myriad philosophical and practical issues that our relatively small group had endlessly debated and never resolved?

I thought about the wisdom of another Aldo Leopold conjecture, written back in 1935: "I suspect there are two categories of judgement which *cannot* be delegated to experts, which every man *must* judge for himself, and on which the intuitive conclusion of the non-expert is perhaps as likely to be correct as that of the professional. One of these is what is right. The other is what is beautiful."

I also realized that there was not now, and most likely never would be, any theoretical or applied silver bullet that could provide the knowledge, resources, and political will to jump-start the restoration of this and most, if not all, of the world's other tragically degraded ecosystems. At least for the foreseeable future, we would be stuck with our inadequate ecological knowledge and technological ability as well as our often factionalized and divisive individual and institutional personalities and perspectives.

Then something shifted in my brain, and suddenly I knew what I would do if I were in charge of restoring this region of the island: I would create a Meta–Intelligent Tinkering "Adopt-an-Acre" Program, in which each semi-independent group of self-sorted people would receive its own parcel of degraded land to restore.

Beyond some commonsense guidelines that everyone could agree on (e.g., no cutting down endangered native trees, planting invasive alien species, or constructing high-end golf courses), there would be no a priori requirement that any group must test some general scientific hypothesis or adhere to a set of standardized and rigorous data collection and monitoring protocols. On the contrary, each group would have the freedom to employ whatever methodologies its members believed would best help them accomplish their particular goals, whether those were, for example, formal scientific research, preservation of endangered species, or ethnobotanic education. At regular time intervals, a democratically elected governing board would evaluate each group and adjust its acreage in accordance with its previous performance, future plans, and general value to the overall restoration program and larger surrounding region.

I believe the community involvement, diversity of approaches, healthy competition and camaraderie, and accountability resulting from this model would foster more and better restoration and greater public involvement, understanding, and support for restoration and conservation in general. If some groups ended up with distinct methodologies and goals, wouldn't that be a lot smarter than putting all our eggs in one basket? Similarly, wouldn't we end up learning more about how to effectively bridge the science-practice gap in particular, and the human-nature interface in general, than we would by implementing a more disciplined program strictly driven by, say, a particular scientific theory or cultural paradigm?

Some would undoubtedly find the heterogeneous checkerboard of acres resulting from this meta–intelligent tinkering model to be inauthentic or at least ecologically compromised. Yet, given the overwhelming and escalating effects of such factors as alien species invasions, functional and actual native species extinctions, and climate change, even the most rigorous and unified restoration programs might not be able to bring back historically "authentic" ecosystems or ecological trajectories, even if we knew what such systems and processes should look like and agreed that they were more valuable and appropriate than other restoration targets.

Moreover, as imperfect as these different "restored" acres might be, most of us would agree that they would be a vast improvement over the ecological and cultural wastelands they would replace. Thus, when it comes to the restoration of highly degraded ecosystems such as tropical dry forests, I would argue that the most important overarching question we can ask is "What are the relative risks and benefits of continuing to do nothing versus doing the best restoration work that we can right now?"

As we continue expanding the scope and scale of our restoration science and practice, perhaps we will eventually find that there are at least some times and places in which the paradigm of formal science fits the target ecological and human landscapes reasonably well and thus can effectively inform and guide the applied restoration work and bridge the science-practice gap. Perhaps we will also find other times and places in which an approach more like intelligent tinkering, or another, presently undiscovered, way of seeing and knowing and doing is more successful and appropriate. My bet is that the enormous challenges created by our planet's extraordinary diversity of human cultures and degraded ecosystems will ultimately require an extraordinary diversity of methodologies and perspectives to enable us to preserve, restore, and reconnect. Yet as people in Hawai'i and throughout the world are increasingly demonstrating, we *can* accomplish these critically important goals when we are willing to put our hearts and minds together.

SELECTED BIBLIOGRAPHY

This bibliography lists the most important and relevant sources I consulted while researching and writing this book. It also cites sources of information, ideas, and quotations I obtained from other works that were not common knowledge. Finally, it suggests a few sources of additional information about some of the topics covered in this book.

Introduction: The Science of Restoration Ecology and the Practice of Ecological Restoration

More about Hawaiian and tropical dry forests

Bruegmann, Marie M. "Hawaii's Dry Forests." *Endangered Species Bulletin* 11 (1996): 26–27.

Bullock, Stephen H., Harold A. Mooney, and Ernesto Medina. *Seasonally Dry Tropical Forests.* New York: Cambridge University Press, 1995.

Cabin, Robert J., Stephen G. Weller, David H. Lorence, Tim W. Flynn, Ann K. Sakai, Darren Sandquist, and Lisa J. Hadway. "Effects of Long-Term Ungulate Exclusion and Recent Alien Species Control on the Preservation and Restoration of a Hawaiian Tropical Dry Forest." *Conservation Biology* 14, no. 2 (April 2000): 439–53.

Janzen, Daniel H. "Tropical Dry Forests: The Most Endangered Major Tropical Ecosystem." In *Biodiversity*, edited by E. O. Wilson, 130–37. Washington, DC: National Academies Press, 1988.

Murphy, Peter G., and Ariel E. Lugo. "Ecology of Tropical Dry Forest." *Annual Review of Ecology and Systematics* 17 (1986): 67–88.

More about extinction, endangerment, and alien species in Hawai'i

Cuddihy, Linda W., and Charles P. Stone. *Alteration of Native Hawaiian Vegetation.* Honolulu: University of Hawai'i, Cooperative National Park Resources Studies Unit, 1990.

Hawaiian Ecosystems at Risk (HEAR) project's website. Last modified October 21, 2010. http://www.hear.org/.

Loope, Lloyd L. "Hawaii and the Pacific Islands." In *Status and Trends of the Nation's Biological Resources*, edited by Michael J. Mac, Paul A. Opler, Catherine E. Puckett Haecker, and Peter D. Doran, 747–77. Reston, VA: US Department of the Interior, US Geological Survey, 1998.

Mehrhoff, Loyal A. "Endangered and Threatened Species." In *Atlas of Hawai'i*, 3rd ed., edited by Sonia P. Juvik and James O. Juvik, 150–53. Honolulu: University of Hawai'i Press, 1998.

Royte, Elizabeth. "On the Brink: Hawaii's Vanishing Species." *National Geographic* 188, no. 3 (September 1995): 2–37.

Rutledge, Daniel T., Christopher A. Lepczyk, Jialong Xie, and Jianguo Liu. "Spatiotemporal Dynamics of Endangered Species Hotspots in the United States." *Conservation Biology* 15, no. 2 (April 2001): 475–87.

Sakai, Ann K., Warren L. Wagner, and Loyal A. Mehrhoff. "Patterns of Endangerment in the Hawaiian Flora." *Systematic Biology* 51, no. 2 (2002): 276–302.

Stone, Charles P., Clifford W. Smith, and J. Timothy Tunison. *Alien Plant Invasions in Native Ecosystems of Hawai'i*. Honolulu: University of Hawai'i, Cooperative National Park Resources Studies Unit, 1992.

Warshauer, F. R. "Alien Species and Threats to Native Ecology." In *Atlas of Hawai'i*, 3rd ed., edited by Sonia P. Juvik and James O. Juvik, 146–49. Honolulu: University of Hawai'i Press, 1998.

Ziegler, Alan C. *Hawaiian Natural History, Ecology, and Evolution*. Honolulu: University of Hawai'i Press, 2002.

More about restoration ecology and ecological restoration

Baldwin, A. Dwight, Jr., Judith de Luce, and Carl Pletsch, eds. *Beyond Preservation: Restoring and Inventing Landscapes*. Minneapolis: University of Minnesota Press, 1994.

Clewell, Andre F., and James Aronson. *Ecological Restoration: Principles, Values, and Structure of an Emerging Profession*. Washington, DC: Island Press, 2007.

Davis, Mark A., and Lawrence B. Slobodkin. "The Science and Values of Restoration Ecology." *Restoration Ecology* 12, no. 1 (March 2004): 1–3.

Egan, Dave, and Evelyn A. Howell. *The Historical Ecology Handbook: A Restorationist's Guide to Reference Ecosystems*. Washington, DC: Island Press, 2001.

Gobster, Paul H., and R. Bruce Hull. *Restoring Nature: Perspectives from the Social Sciences and Humanities*. Washington, DC: Island Press, 2000.

Higgs, Eric S. *Nature by Design: People, Natural Process, and Ecological Restoration*. Cambridge, MA: MIT Press, 2003.

Hobbs, Richard J. "The Future of 'Restoration Ecology': Challenges and Opportunities." *Restoration Ecology* 13, no. 2 (June 2005): 239–41.

Jordan, William R., III. *The Sunflower Forest: Ecological Restoration and the New Communion with Nature*. Berkeley: University of California Press, 2003.

Katz, Eric. "The Problem of Ecological Restoration." *Environmental Ethics* 18 (Summer 1996): 222–24.

Perrow, Martin R., and Anthony J. Davy, eds. *Handbook of Ecological Restoration*. 2 vols. Vol. 1, *Principles of Restoration*. Vol. 2, *Restoration in Practice*. Cambridge: Cambridge University Press, 2002.

Society for Ecological Restoration's website. Accessed October 30, 2010. http://www
.ser.org/.

Temperton, Vicky M., Richard J. Hobbs, Tim Nuttle, and Stefan Halle, eds. *Assembly
Rules and Restoration Ecology: Bridging the Gap between Theory and Practice.*
Washington, DC: Island Press, 2004.

Throop, William, ed. *Environmental Restoration: Ethics, Theory, and Practice.* New
York: Humanity Books, 2000.

Van Andel, Jelte, and James Aronson, eds. *Restoration Ecology: The New Frontier.*
Malden, MA: Blackwell Science, 2006.

Warshauer, F. R. "Alien Species and Threats to Native Ecology." In *Atlas of Hawai'i,*
3rd ed., edited by Sonia P. Juvik and James O. Juvik, 146–49. Honolulu: University
of Hawai'i Press, 1998.

Winterhalder, Keith, Andre F. Clewell, and James Aronson. "Values and Science in
Ecological Restoration—a Response to Davis and Slobodkin." *Restoration Ecology*
12, no. 1 (March 2004): 4–7.

Leopold's "alone in a world of wounds" quote

Leopold, Aldo. *A Sand County Almanac.* Oxford: Oxford University Press, 1966.

Restoration ecology as the "acid test" of academic ecology

Bradshaw, Anthony D. "Restoration: The Acid Test for Ecology." In *Restoration Ecol-
ogy: A Synthetic Approach to Ecological Research*, edited by William R. Jordan III,
Michael E. Gilpin, and John D. Aber, 23–29. Cambridge: Cambridge University
Press, 1987.

Leopold's "intelligent tinkering" quote

Leopold, Aldo. *A Sand County Almanac.* Oxford: Oxford University Press, 1966.

*Leopold's "land organism" and "senseless barrier between science
and art" quotes*

Knight, Richard L., and Suzanne Riedel, eds. *Aldo Leopold and the Ecological Conscience.* Oxford: Oxford University Press, 2002.

Leopold's "doctor sick land" quote

Meine, Curt, and Richard L. Knight, eds. *The Essential Aldo Leopold.* Madison: University of Wisconsin Press, 1999.

More about Leopold's life, work, and ideas

Knight, Richard L., and Suzanne Riedel, eds. *Aldo Leopold and the Ecological Conscience.* Oxford: Oxford University Press, 2002.

Meine, Curt. *Aldo Leopold: His Life and Work.* Madison: University of Wisconsin
Press, 1989.

Meine, Curt, and Richard L. Knight, eds. *The Essential Aldo Leopold*. Madison: University of Wisconsin Press, 1999.

Chapter 1. Tropical Dry Forests: Land of the Living Dead

History of alien plant species invasions and habitat destruction in Hawai'i

Cuddihy, Linda W., and Charles P. Stone. *Alteration of Native Hawaiian Vegetation*. Honolulu: University of Hawai'i, Cooperative National Park Resources Studies Unit, 1990.
Gagne, Wayne C., and Linda W. Cuddihy. "Vegetation." In *Manual of the Flowering Plants of Hawai'i*, rev. ed., edited by Warren L. Wagner, Derral R. Herbst, and S. H. Sohmer, 45–114. Honolulu: University of Hawai'i Press, Bishop Museum Press, 1999.

Hawaiian ecology, evolution, natural history, conservation, and biogeography

Berger, Andrew J. *Hawaiian Birdlife*, 2nd ed. Honolulu: University of Hawai'i Press, 1981.
Carlquist, Sherwin. *Hawaii: A Natural History*. Garden City, NY: Natural History Press for the American Museum of Natural History, 1970.
Gagne, Wayne C. "Conservation Priorities in Hawaiian Natural Systems." *BioScience* 38, no. 4 (1988): 264–71.
Hawai'i Audubon Society. *Hawaii's Birds*. Honolulu: Hawaii Audubon Society, 1993.
Hawai'i Conservation Alliance's website. Accessed October 30, 2010. http://hawaiiconservation.org/.
Howarth, Francis G., and William P. Mull. *Hawaiian Insects and Their Kin*. Honolulu: University of Hawai'i Press, 1992.
Howarth, F. G., S. H. Sohmer, and W. D. Duckworth. "Hawaiian Natural History and Conservation Efforts: What's Left Is Worth Saving." *BioScience* 38, no. 4 (1988): 232–37.
Juvik, Sonia P., and James O. Juvik, eds. *Atlas of Hawai'i*, 3rd ed. Honolulu: University of Hawai'i Press, 1998.
Kay, E. Allison, ed. *A Natural History of the Hawaiian Islands: Selected Readings II*. Honolulu: University of Hawai'i Press, 1994.
Liittschwager, David, and Susan Middleton. *Remains of a Rainbow: Rare Plants and Animals of Hawai'i*. Washington, DC: National Geographic Society, 2001.
Luna, Tara. "Native Plant Restoration on Hawai'i." *Native Plants* 4, no. 1 (Spring 2003): 22–36.
Pratt, H. Douglas, Phillip L. Bruner, and Delwyn G. Berrett. *A Field Guide to the Birds of Hawaii and the Tropical Pacific*. Princeton, NJ: Princeton University Press, 1987.
Scott, J. Michael, Cameron B. Kepler, Charles van Riper III, and Stewart I. Fefer. "Conservation of Hawaii's Vanishing Avifauna." *BioScience* 38, no. 4 (1988): 238–53.
Stone, Charles P., and Danielle B. Stone, eds. *Conservation Biology in Hawai'i*. Honolulu: University of Hawai'i, Cooperative National Park Resources Studies Unit, 1989.

Wagner, Warren L., and V. A. Funk, eds. *Hawaiian Biogeography: Evolution on a Hot Spot Archipelago*. Washington, DC: Smithsonian Institution Press, 1995.

Ziegler, Alan C. *Hawaiian Natural History, Ecology, and Evolution*. Honolulu: University of Hawai'i Press, 2002.

Fountain grass invasion on the island of Hawai'i

Jacobi, James D., and Fredrick R. Warshauer. "Distribution of Six Alien Plant Species in Upland Habitats on the Island of Hawai'i." In *Alien Plant Invasions in Native Ecosystems of Hawai'i*, edited by Charles P. Stone, Clifford W. Smith, and J. Timothy Tunison, 155–88. Honolulu: University of Hawai'i, Cooperative National Park Resources Studies Unit, 1992.

Wagner, Warren L., Derral R. Herbst, and S. H. Sohmer. *Manual of the Flowering Plants of Hawai'i*, rev. ed. Honolulu: University of Hawai'i Press and Bishop Museum Press, 1999.

Population biology of invasive species

Sakai, Ann K., Fred W. Allendorf, Jodie S. Holt, David M. Lodge, Jane Molofsky, Kimberly A. With, Syndallas Baughman, Robert J. Cabin, Joel E. Cohen, Norman C. Ellstrand, David E. McCauley, Pamela O'Neil, Ingrid M. Parker, John N. Thompson, and Stephen G. Weller. "The Population Biology of Invasive Species." *Annual Review of Ecology and Systematics* 32 (November 2001): 305–32.

Hawaiian ethnobotany

Abbott, Isabella A. *La'au Hawai'i: Traditional Hawaiian Uses of Plants*. Honolulu: Bishop Museum Press, 1992.

Kane, Herb K. *Ancient Hawai'i*. Captain Cook, HI: Kawainui Press, 1997.

Liittschwager, David, and Susan Middleton. *Remains of a Rainbow: Rare Plants and Animals of Hawai'i*. Washington, DC: National Geographic Society, 2001.

Merrill, Nancy. *Self-Guided Tour of Limahuli Garden*. Ha'ena, Kaua'i, HI: National Tropical Botanical Garden, 1998.

Sohmer, S. H., and R. Gustafson. *Plants and Flowers of Hawai'i*. Honolulu: University of Hawai'i Press, 1987.

Mammals of Hawai'i

Tomich, P. Quentin. *Mammals in Hawai'i: A Synopsis and Notational Bibliography*. Honolulu: Bishop Museum Press, 1969.

Prehistoric ecology of Hawai'i

Burney, David A., Helen F. James, Lida Pigott Burney, Storrs L. Olson, William Kikuchi, Warren L. Wagner, Mara Burney, Deirdre McCloskey, Delores Kikuchi, Frederick V. Grady, Reginald Gage II, and Robert Nishek. "Fossil Evidence for a

Diverse Biota from Kaua'i and Its Transformation since Human Arrival." *Ecological Monographs* 71, no. 4 (2001): 615–41.

Kane, Herb K. *Ancient Hawai'i*. Captain Cook, HI: Kawainui Press, 1997.

Kirch, Patrick V., and Terry L. Hunt, eds. *Historical Ecology in the Pacific Islands*. New Haven, CT: Yale University Press, 1997.

Ziegler, Alan C. *Hawaiian Natural History, Ecology, and Evolution*. Honolulu: University of Hawai'i Press, 2002.

Joseph Rock's book about Hawai'i's tree flora

Rock, Joseph F. *The Indigenous Trees of the Hawaiian Islands*. Honolulu: T. H., 1913; reprint, Lāwa'i, Kaua'i, HI: Pacific Tropical Botanical Garden and Charles E. Tuttle, 1974.

More about dry forests, the Ka'upulehu Dry Forest Preserve, and the North Kona Dryland Forest Working Group

Hawai'i Forest Industry Association. "Hawaii's Dryland Forests: Can They Be Restored?" Accessed October 30, 2010. http://www.hawaiiforest.org/reports/dryland.html.

Ka'ahahui 'O Ka Nāhelehele. "Nāhelehele's Hawai'i Dryland Forest." Accessed October 30, 2010. http://www.drylandforest.org/.

Chapter 2. Let's See Action! Planning and Implementing a Research and Restoration Program

Geologic overview of the Hawaiian Islands

Clague, David A. "Geology." In *Atlas of Hawai'i*, 3rd ed., edited by Sonia P. Juvik and James O. Juvik, 37–46. Honolulu: University of Hawai'i Press, 1998.

Walker, George. "Geology." In *Manual of the Flowering Plants of Hawai'i*, rev. ed., edited by Warren L. Wagner, Derral R. Herbst, and S. H. Sohmer, 21–35. Honolulu: University of Hawai'i Press and Bishop Museum Press, 1999.

Ziegler, Alan C. *Hawaiian Natural History, Ecology, and Evolution*. Honolulu: University of Hawai'i Press, 2002.

Hawaiian founder events; silversword genetics

Baldwin, Bruce G., and Robert H. Robichaux. "Historical Biogeography and Ecology of the Hawaiian Silversword Alliance (Asteraceae): New Molecular Phylogenetic Perspectives." In *Hawaiian Biogeography: Evolution on a Hot Spot Archipelago*, edited by Warren L. Wagner and V. A. Funk, 259–87. Washington, DC: Smithsonian Institution Press, 1995.

Carson, H. L. "Evolution." In *Atlas of Hawai'i*, 3rd ed., edited by Sonia P. Juvik and James O. Juvik, 107–10. Honolulu: University of Hawai'i Press, 1998.

Flora and ecology of Mahanaloa Gulch

Weller, Stephen G., Robert J. Cabin, David H. Lorence, Steven Perlman, Ken Wood, Timothy Flynn, and Ann K. Sakai. "Alien Plant Invasions, Introduced Ungulates, and Alternative States in a Mesic Forest in Hawaii." *Restoration Ecology,* in press. doi:10.1111/j.1526-100X.2009.00635.x.

National Tropical Botanical Garden

Haapoja, Margaret A. "Nature's Keeper." *Aloha,* September–October 1996, 30–35.
Milius, Susan. "Emergency Gardening: Labs Step In to Help Conserve the Rarest Plants on Earth." *Science News* 164, no. 6 (August 2003): 88–90.
National Tropical Botanical Garden's website. Accessed October 30, 2010. http://www.ntbg.org/.

First Kaʻupulehu outplanting

Cabin, Robert J. "Outplanting at Kaʻupulehu: Story behind the Trees." *Woods* 7 (1999): 1–3.

Chapter 3. Now What? Responding to Nature's Response

Colubrina, kauila, *and* Alphitonia *taxonomy and uses*

Abbott, Isabella A. *Laʻau Hawaiʻi: Traditional Hawaiian Uses of Plants.* Honolulu: Bishop Museum Press, 1992.
Wagner, Warren L., Derral R. Herbst, and S. H. Sohmer. *Manual of the Flowering Plants of Hawaiʻi,* rev. ed. Honolulu: University of Hawaiʻi Press and Bishop Museum Press, 1999.

More about the Kaʻupulehu research discussed in this chapter

Cabin, Robert J., Stephen G. Weller, David H. Lorence, Tim W. Flynn, Ann K. Sakai, Darren Sandquist, and Lisa J. Hadway. "Effects of Long-Term Ungulate Exclusion and Recent Alien Species Control on the Preservation and Restoration of a Hawaiian Tropical Dry Forest." *Conservation Biology* 14, no. 2 (April 2000): 439–53.

Chapter 4. Writing It Up: The Art and Importance of Science Papers

The scientific paper discussed in this chapter

Cabin, Robert J., Stephen G. Weller, David H. Lorence, Tim W. Flynn, Ann K. Sakai, Darren Sandquist, and Lisa J. Hadway. "Effects of Long-Term Ungulate Exclusion and Recent Alien Species Control on the Preservation and Restoration of a Hawaiian Tropical Dry Forest." *Conservation Biology* 14, no. 2 (April 2000): 439–53.

Chapter 5. Scaling Up: Micro to Macro Science and Practice

Management of Puʻu Waʻawaʻa Ranch

Tummons, Patricia. "Puʻuwaʻawaʻa at the Crossroads." *Environment Hawaiʻi* 11, no. 3 (September 2000).

Rock's "richest floral section of any in the whole Territory" quote

Rock, Joseph F. *The Indigenous Trees of the Hawaiian Islands.* Honolulu: T. H., 1913; reprint, Lāwaʻi, Kauaʻi, HI: Pacific Tropical Botanical Garden and Charles E. Tuttle, 1974.

The seeding experiment

Cabin, Robert J., Stephen G. Weller, David H. Lorence, Susan Cordell, and Lisa J. Hadway. "Effects of Microsite, Water, Weeding, and Direct Seeding on the Regeneration of Native and Alien Species within a Hawaiian Dry Forest Preserve." *Biological Conservation* 104, no. 2 (April 2002): 181–90.

Chapter 6. Shall We Dance? The Trade-Offs of Science-Practice Collaborations and Community-Driven Restoration

The Big Experiment

Cabin, Robert J., Stephen G. Weller, David H. Lorence, Susan Cordell, Lisa J. Hadway, Rebecca Montgomery, Don Goo, and Alan Urakami. "Effects of Light, Alien Grass, and Native Species Additions on Hawaiian Dry Forest Restoration." *Ecological Applications* 12, no. 6 (2002): 1595–1610.

More about the Kaʻupulehu research and restoration programs

Allen, William. "At Kaʻupulehu, a Dryland Forest Is Lovingly Restored." *Environment Hawaiʻi* 11, no. 3 (September 2000): 10–11.
——. "Restoring Hawaii's Dry Forests." *BioScience* 50, no. 12 (December 2000): 1037–41.
Cabin, Robert J., Stephen G. Weller, David H. Lorence, and Lisa J. Hadway. "Restoring Tropical Dry Forests with Direct Seeding: The Effects of Light, Water, and Weeding (Hawaii)." *Ecological Restoration* 17, no. 4 (Winter 1999): 237–38.
Cordell, Susan, Robert J. Cabin, and Lisa J. Hadway. "Physiological Ecology of Native and Alien Dry Forest Shrubs in Hawaii." *Biological Invasions* 4, no. 4 (2002): 387–96.
Cordell, Susan, Robert J. Cabin, Stephen G. Weller, and David H. Lorence. "Simple and Cost-Effective Methods Control Fountain Grass in Dry Forests (Hawaii)." *Ecological Restoration* 20 (2002): 139–40.
Cordell, Susan, and Darren R. Sandquist. "The Impact of an Invasive African Bunchgrass (*Pennisetum setaceum*) on Water Availability and Productivity of Canopy

Trees within a Tropical Dry Forest in Hawaii." *Functional Ecology* 22, no. 6 (2008): 1008–17.

Litton, Creighton M., Darren R. Sandquist, and Susan Cordell. "Effects of Non-native Grass Invasion on Aboveground Carbon Pools and Tree Population Structure in a Tropical Dry Forest of Hawaii." *Forest Ecology and Management* 231 (2006): 105–13.

———. "A Non-native Invasive Grass Increases Soil Carbon Flux in a Hawaiian Tropical Dry Forest." *Global Change Biology* 14, no. 4 (April 2008): 726–39.

Meadows, Robin. "RX for Hawaii's Dry Forests: No Cows and Lots of Hard Work." *Conservation Biology in Practice* 1 (2000): 6–7.

Sandquist, Darren R., and Susan Cordell. "Functional Diversity of Carbon-Gain, Water-Use, and Leaf-Allocation Traits in Trees of a Threatened Lowland Dry Forest in Hawaii." *American Journal of Botany* 94 (2007): 1459–69.

Thaxton, Jarrod M., Colleen Cole, Susan Cordell, Robert J. Cabin, Darren R. Sandquist, and Creighton M. Litton. "Native Species Regeneration following Ungulate Exclusion and Nonnative Grass Removal in a Remnant Hawaiian Dry Forest." *Pacific Science* 64, no. 4 (2010): 533–44.

Chapter 7. The Science-Practice Gap

Articles from which parts of this chapter were adapted

Cabin, Robert J. "Science and Restoration under a Big, Demon Haunted Tent: Reply to Giardina et al. (2007)." *Restoration Ecology* 15, no. 3 (September 2007): 377–81.

———. "Science-Driven Restoration: A Square Grid on a Round Earth?" *Restoration Ecology* 15, no. 1 (March 2007): 1–7.

Makua Military Reservation restoration implementation plan

US Army Garrison, Hawaii, Directorate of Public Works, Environmental Division. "Addendum to the Implementation Plan, Makua Military Reservation, Island of Oʻahu." January 2005. http://manoa.hawaii.edu/hpicesu/DPW/2003_MIP/Add /001.pdf.

———. "2005 Status Report: Makua Implementation Plan, Island of Oʻahu." September 2005. http://manoa.hawaii.edu/hpicesu/DPW/2005_MIP/TOC.pdf.

Kōkeʻe Resource Conservation Program

Cassel, Katie, and Ellen Coulombe. "A Forest Worth Saving on Kauai." *Agriculture Hawaii*, January–March 2002, 13.

Conrow, Joan. "Lend a Helping Hand." *Hawaiʻi*, September–October 2003, 26–28.

Fullard-Leo, Betty. "Volunteer Vacations: Working for the Pure Pleasure of It." *Spirit of Aloha*, May–June 2001.

Garden Island Resource Conservation and Development. "Kōkeʻe Resource Conservation Program." Accessed October 30, 2010. http://www.krcp.org/.

Uninformed gut feelings

Pohlon, E., C. Augsperger, U. Risse-Buhl, J. Arle, M. Willkomm, S. Halle, and K. Küsel. 2007. Querying the obvious: lessons from a degraded stream. *Restoration Ecology* 15:312–316.

Dictionary definitions of science

Brown, Lesley, ed. *The New Shorter Oxford English Dictionary*. Oxford: Clarendon Press, 2003.

Eminent ecological historian's definition of science

Merchant, Carolyn. *Ecological Revolutions*. Chapel Hill: University of North Carolina Press, 1989.

SER's definition of restoration ecology

Society for Ecological Restoration (SER) Science and Policy Working Group. *The SER International Primer on Ecological Restoration*. Tucson, AZ: Society for Ecological Restoration, 2004. Online at http://www.ser.org/.

Hawaiian scientists' paraphrasing of SER's definition of restoration ecology

Giardina, Christian P., Creighton M. Litton, Jarrod M. Thaxton, Susan Cordell, Lisa J. Hadway, and Darren R. Sandquist. "Science Driven Restoration: A Candle in a Demon Haunted World—Response to Cabin (2007)." *Restoration Ecology* 15, no. 2 (June 2007): 171–76.

Scientific authoritarianism

Higgs, E. 2005. The two-culture problem: ecological restoration and the integration of knowledge. *Restoration Ecology* 13:159–164.

Quote about the response of Hawaiian ecosystems to removal of feral pigs and goats

Stone, Charles P., Linda W. Cuddihy, and J. Timothy Tunison. "Responses of Hawaiian Ecosystems to Removal of Feral Pigs and Goats." In *Alien Plant Invasions in Native Ecosystems of Hawai'i*, edited by Charles P. Stone, Clifford W. Smith, and J. Timothy Tunison, 666–704. Honolulu: University of Hawai'i, Cooperative National Park Resources Studies Unit, 1992.

David Ehrenfeld quotes

Ehrenfeld, David. "War and Peace and Conservation Biology." *Conservation Biology* 14, no. 1 (February 2000): 105–12.

Chapter 8. Bridging the Science-Practice Gap

Article from which parts of this chapter were adapted

Cabin, Robert J., Andre Clewell, Mrill Ingram, Tein McDonald, and Vicky Temperton. "Bridging Restoration Science and Practice: Results and Analysis of a Survey from the 2009 Society for Ecological Restoration International Meeting." *Restoration Ecology* 18, no. 6 (November 2010): 783–88.

Quote about restoration scientists and practitioners working together

Jordan, William R., III, Michael E. Gilpin, and John D. Aber, eds. *Restoration Ecology: A Synthetic Approach to Ecological Research*. Cambridge: Cambridge University Press, 1987.

SER's Global Restoration Network

Society for Ecological Restoration. "Global Restoration Network." Accessed June 9, 2010. http://www.globalrestorationnetwork.org/.

Quote from the prominent ecologist's review of the factions in academic ecology

McIntosh, Robert P. "Pluralism in Ecology." *Annual Review of Ecology and Systematics* 18 (1987): 321–41.

SER's mission statement

Society for Ecological Restoration. "About SER." Accessed October 31, 2010. http://www.ser.org/about.asp.

Source of the physics envy *term*

Egler, Frank E. "'Physics Envy' in Ecology." *Bulletin of the Ecological Society of America* 67, no. 3 (September 1986): 233–35.

More about alternative research models

Cabin, Robert J. "Science-Driven Restoration: A Square Grid on a Round Earth?" *Restoration Ecology* 15, no. 1 (March 2007): 1–7.

Quote about the difficulty of resolving political disputes with science

Sarewitz, Daniel. "Liberating Science from Politics." *American Scientist* 94, no. 3 (May–June 2006): 104–6.

Mikael Stenmark quotes

Stenmark, Mikael. *Scientism: Science, Ethics, and Religion*. London: Ashgate Publishing, 2001.

More about the "Restoration Ecology Extension Service" idea

Cabin, Robert J., Andre Clewell, Mrill Ingram, Tein McDonald, and Vicky Temperton. "Bridging Restoration Science and Practice: Results and Analysis of a Survey from the 2009 Society for Ecological Restoration International Meeting." *Restoration Ecology* 18, no. 6 (November 2010): 783–88.

Ken Wood's discovery of Kanaloa

Lorence, David H., and Kenneth R. Wood. "*Kanaloa*, a New Genus of Fabaceae (Mimosoideae) from Hawaii." *Novon* 4, no. 2 (Summer 1994): 137–45.

Selected publications from my seed bank dissertation research

Cabin, Robert J. "Genetic Comparisons of Seed Bank and Seedling Populations of a Perennial Desert Mustard, *Lesquerella fendleri*." *Evolution* 50, no. 5 (October 1996): 1830–41.
Cabin, Robert J., and Diane L. Marshall. "The Demographic Role of Soil Seed Banks. I. Spatial and Temporal Comparisons of Below- and Above-Ground Populations of the Desert Mustard *Lesquerella fendleri*." *Journal of Ecology* 88, no. 2 (April 2000): 283–92.
Cabin, Robert J., Randall J. Mitchell, and Diane L. Marshall. "Do Surface Plant and Soil Seed Bank Populations Differ Genetically? A Multipopulation Study of the Desert Mustard *Lesquerella fendleri* (Brassicaceae)." *American Journal of Botany* 85, no. 9 (1998): 1098–1109.
Evans, Ann S., and Robert J. Cabin. "Can Dormancy Affect the Evolution of Post-germination Traits? The Case of *Lesquerella fendleri*." *Ecology* 76, no. 2 (March 1995): 344–56.

Chapter 9. Intelligent Tinkering

Articles from which parts of this chapter were adapted

Cabin, Robert J. "Science and Restoration under a Big, Demon Haunted Tent: Reply to Giardina et al. (2007)." *Restoration Ecology* 15, no. 3 (September 2007): 377–81.
———. "Science-Driven Restoration: A Square Grid on a Round Earth?" *Restoration Ecology* 15, no. 1 (March 2007): 1–7.

Source of the phrase "remains of a rainbow"

Liittschwager, David, and Susan Middleton. *Remains of a Rainbow: Rare Plants and Animals of Hawai'i*. Washington, DC: National Geographic Society, 2001.

Leopold's "fusion point of science and the land community" quote

Knight, Richard L., and Suzanne Riedel, eds. *Aldo Leopold and the Ecological Conscience*. Oxford: Oxford University Press, 2002.

Leopold's "reversal of specialization" and "more and more about less and less" quotes

Meine, Curt, and Richard L. Knight, eds. *The Essential Aldo Leopold*. Madison: University of Wisconsin Press, 1999.

More about the Hakalau restoration program

Friends of Hakalau Forest National Wildlife Refuge's website. Accessed October 30, 2010. http://www.friendsofhakalauforest.org/.
Furukawa, George. "The Mother of the Rainforest." *American Forests* 107 (Winter 2002): 40–43.
Jeffrey, Jack, and Baron Horiuchi. "Tree Planting at Hakalau Forest National Wildlife Refuge." *Native Plants* 4, no. 1 (Spring 2003): 30–31.
Levy, Sharon. "Empty Nest Syndrome." *OnEarth* 27 (Summer 2005): 27–31.
Scowcroft, Paul G., and Jack Jeffrey. "Potential Significance of Frost, Topographic Relief, and *Acacia koa* Stands to Restoration of Mesic Hawaiian Forests on Abandoned Rangeland." *Forest Ecology and Management* 114, nos. 2–3 (February 1999): 447–58.
Scowcroft, Paul G., Frederick C. Meinzer, Guillermo Goldstein, Peter J. Melcher, and Jack Jeffrey. "Moderating Night Radiative Cooling Reduces Frost Damage to *Metrosideros polymorpha* Seedlings Used for Forest Restoration in Hawaii." *Restoration Ecology* 8, no. 2 (June 2000): 161–69.
US Fish and Wildlife Service. "Hakalau Forest National Wildlife Refuge." Last updated October 18, 2010. http://www.fws.gov/hakalauforest/.

Art Medeiros's description of the 'awa ceremony; more about the Auwahi restoration program

Medeiros, A. C. "The Pū'Olē'Olē Blows and 'Awa Is Poured." In *Wao Akua: Sacred Source of Life*, 113–19. Honolulu: State of Hawai'i, Department of Land and Natural Resources, 2003.
Medeiros, A. C., C. F. Davenport, and C. G. Chimera. "Auwahi: Ethnobotany of a Hawaiian Dryland Forest." Accessed October 30, 2010. http://www.hear.org /naturalareas/auwahi/ethnobotany_of_auwahi.pdf.
Medeiros, Arthur C., and Erica von Allmen. "Restoration of Native Hawaiian Dryland Forest at Auwahi, Maui." Accessed October 30, 2010. http://biology.usgs.gov/pierc /Pollution_&_Ecological_Restoration/Dryland_restoration.pdf.
Starr, Forest, and Kim Starr. "Natural Areas of Hawaii." Last modified January 13, 2008. http://www.hear.org/naturalareas/auwahi/.

More about the Limahuli restoration program

Blaine, Jessica MacMurray. "Hawaii's Limahuli Garden and Preserve." *Forest* (Winter 2004): 39–41.

McCormick, Kathleen. "Cultivating the Genuine Kauai." *New York Times*, May 26, 1996, 10–11, 15.

National Tropical Botanical Garden. "Limahuli Garden and Preserve." Accessed October 30, 2010. http://www.ntbg.org/gardens/limahuli.php.

Leopold's "I suspect there are two categories of judgement" quote

Meine, Curt, and Richard L. Knight, eds. *The Essential Aldo Leopold.* Madison: University of Wisconsin Press, 1999.

INDEX

'A'ā (lava type), 2, 18
'A'ali'i (Dodonaea viscosa), 182, 183
Acacia koa, 176–78
Adaptation, local, vs. genetic diversity, 61–65
Adaptive management protocols, 158
African fountain grass. See Fountain grass (Pennisetum setaceum)
Ahupua'a of Ka'upulehu, 25
Ahupua'a of Kuki'o, 25
'Aiea (Nothocestrum breviflorum), 30, 129–30
Alahe'e (Psydrax odorata), 3
'Alalā (Hawaiian crow), 26
Aleutian Islands, 35
Alien animal species, 3
Alien nurse plant approach, 78–79
Alien plant species
    aerial seeding in 1950s, 23–24
    in fountain grass control project, 76
    on the Hawaiian islands, 7
    invasion facilitated by exotic grazing animals, 14–15
    lag phase in colonization by, 17
    manipulation of, and native plant regeneration, 78–79
    See also Fountain grass (Pennisetum setaceum); Lantana (Lantana camara)
Allerton, John, 39
Allerton, Robert, 39

Allerton Garden, NTBG, 39–40
Alphitonia ponderosa (kauila), 61–63, 65, 75
Alyxia oliviformis (maile), 183
American Scientist (periodical), 160
Applied scientists, 163, 173
Argemone glauca (Hawaiian poppy), 103
Auwahi, Maui, xix (map), 179–85
'Awa (Piper methysticum), 181

Beck, Andrea, 66–67, 68–70, 126
Bender, Dave, 186
The Big Experiment
    botanical survey, 102
    constraining factors, 107–8
    experimental blocks, 108, 110–12
    fountain grass control treatments, 111–12, 119
    launch day, 109–14
    launch day plus five months, 117–18
    native plant regeneration, 102–3
    native species selection, 110
    seeding, 114
    volunteers, 112–14
Biocontrol agents, in fountain grass control, 104
Biodiversity, ecological triage vs. restoration and preservation of, 7, 64
Biological endemism and richness, on Hawai'i compared to Kaua'i, 36
Bird extinctions, 7

207

THE SCIENCE AND PRACTICE
OF ECOLOGICAL RESTORATION

Wildlife Restoration: Techniques for Habitat Analysis and Animal Monitoring, by Michael L. Morrison

Ecological Restoration of Southwestern Ponderosa Pine Forests, edited by Peter Friederici, Ecological Restoration Institute at Northern Arizona University

Ex Situ Plant Conservation: Supporting Species Survival in the Wild, edited by Edward O. Guerrant Jr., Kayri Havens, and Mike Maunder

Great Basin Riparian Ecosystems: Ecology, Management, and Restoration, edited by Jeanne C. Chambers and Jerry R. Miller

Assembly Rules and Restoration Ecology: Bridging the Gap between Theory and Practice, edited by Vicky M. Temperton, Richard J. Hobbs, Tim Nuttle, and Stefan Halle

The Tallgrass Restoration Handbook: For Prairies, Savannas, and Woodlands, edited by Stephen Packard and Cornelia F. Mutel

The Historical Ecology Handbook: A Restorationist's Guide to Reference Ecosystems, edited by Dave Egan and Evelyn A. Howell

Foundations of Restoration Ecology: The Science and Practice of Ecological Restoration, edited by Donald A. Falk, Margaret A. Palmer, and Joy B. Zedler

Restoring the Pacific Northwest: The Art and Science of Ecological Restoration in Cascadia, edited by Dean Apostol and Marcia Sinclair